Mammals as Pests

Mammals as Pests

EDITED BY

R.J. PUTMAN

on behalf of
The Mammal Society

London New York
CHAPMAN AND HALL

First published in 1989 by Chapman and Hall Ltd
11 New Fetter Lane, London EC4P 4EE

Published in the USA by Chapman and Hall
29 West 35th Street, New York NY 10001

Typeset in 10½ on 12pt Sabon by
EJS Chemical Composition, Bath

Printed in Great Britain by St Edmundsbury Press,
Bury St Edmunds, Suffolk

ISBN 0 412 32590 X

British Library Cataloguing in Publication Data

Mammals as pests.
1. Pests: Mammals. Control measures
I. Putman, R.J. (Roderick J.).
632′69

ISBN 0-412-32590-X

Library of Congress Cataloging in Publication Data

Mammals as pests/edited by R.J. Putman
on behalf of the Mammal Society.
p. cm.
Based on papers presented at symposium organized by the Mammal Society, held at London, 1987.
Bibliography: p.
Includes index.
ISBN 0-412-32590-X
1. Mammal pests—Congresses.
2. Mammal pests— Control-Congresses.
I. Putman, R.J. (Rory J.), 1950– . II. Mammal Society
SB993.5.M36 1989
628.9′69—dc 19 88-31496 CIP

Contents

Contributors

Sheila S. Anderson Sea Mammal Research Unit, Natural Environment Research Council, High Cross, Madingley Road, Cambridge CB3 0ET, UK

Simon J. Baker Coypu Research Laboratory, MAFF, Jupiter Road, Norwich NR6 6SP, UK

Brian Boag 6 Park Road, Invergowrie, Dundee DD2 5AH, UK

Nigel Bonner British Antarctic Survey, Natural Environment Research Council, High Cross, Madingley Road, Cambridge CB3 0ET, UK

W. Graham Christie ADAS Worplesdon Laboratory, Ministry of Agriculture, Fisheries and Food, Tangley Place, Worplesdon, Guildford GU3 3LQ, UK

David P. Cowan ADAS Worplesdon Laboratory, Ministry of Agriculture, Fisheries and Food, Tangley Place, Worplesdon, Guildford GU3 3LQ, UK

Michael J. Crawley Department of Pure and Applied Biology, Imperial College, Silwood Park, Ascot, Berks, UK

Nigel Dunstone Department of Zoology, University of Durham, Science Laboratories, South Road, Durham DH1 3LE, UK

Martyn L. Gorman University of Aberdeen, Culterty Field Station, Newburgh, Ellon AB4 0AA, UK

Morris L. Gosling Coypu Research Laboratory, MAFF, Jupiter Road, Norwich NR6 6SP, UK

John Gurnell School of Biological Sciences, Queen Mary College, Mile End Road, London E1 4NS, UK

Mark J. Hannan Department of Agricultural Zoology and Genetics, University College, Belfield, Dublin 4, Ireland

Anthony R. Hardy ADAS Worplesdon Laboratory, Ministry of Agriculture, Fisheries and Food, Tangley Place, Worplesdon, Guildford GU3 3LQ, UK

Stephen Harris Department of Zoology, University of Bristol, Woodland Road, Bristol BS8 1UG, UK

Michael D.K. Harrison Agricultural Development and Advisory Service, Ministry of Agriculture, Fisheries and Food, Burghill Road, Westbury-on-Trym, Bristol BS10 6NJ, UK

John Harwood Sea Mammal Research Unit, Natural Environment Research Council, High Cross, Madingley Road, Cambridge CB3 0ET, UK

Mark Ireland Department of Zoology, University of Durham, Science Laboratories, South Road, Durham DH1 3LE, UK

Robert E. Kenward Institute of Terrestrial Ecology, Furzebrook Research Station, Wareham, Dorset BH20 5AS, UK

Allan B. Lazarus ADAS Tolworth Laboratory, Hook Rise South, Tolworth, Surbiton, Surrey KT6 7NF, UK

Jens Lodal Statens Skadedyrlaboratorium (Danish Pest Infestation Laboratory), Skovbrynet 14, DK 2800 Lyngby, Denmark

Mogens Lund Statens Skadedyrlaboratorium (Danish Pest Infestation Laboratory), Skovbrynet 14, DK 2800 Lyngby, Denmark

Peter Lüps Naturhistorisches Museum, Bernastrasse 15, CH-3005 Berne, Switzerland

Shali Lycett School of Biology, University of Sussex, Brighton BN1 9QG, UK

André Meylan Station fédérale de recherches agronomiques de Changins, CH-1260 Nyon, Switzerland

Peter N. Neville 1, Coombe House, Hartland Road, Addlestone, Weybridge, Surrey KT15 1JU, UK

Hans-Joachim Pelz Federal Biological Research Center for Agriculture and Forestry, Institute for Nematology and Vertebrate Research, Toppheideweg 88, D-4400 Munster, FRG

John H. Prime Sea Mammal Research Unit, Natural Environment Research Council, High Cross, Madingley Road, Cambridge CB3 0ET, UK

Rory J. Putman Department of Biology, Building 44, The University, Southampton SO9 5NH, UK

Philip R. Ratcliffe Wildlife and Conservation Research Branch, Forestry Commission, Forest Research Station, Alice Holt Lodge, Wrecclesham, Farnham, Surrey GU10 4LH, UK

Christopher G.J. Richards ICI Agrochemicals, Fernhurst, Haslemere, Surrey GU27 3JE, UK
Timothy J. Roper School of Biology, University of Sussex, Brighton BN1 9QG, UK
Graham C. Smith Department of Zoology, University of Bristol, Woodland Road, Bristol BS8 1UG, UK
Brian W. Staines Institute of Terrestrial Ecology, Hill of Brathens, Banchory, Kincardineshire AB3 4BY, UK
R. David Stone Institut de Zoologie et d'écologie animale, Université de Lausanne, Bâtiment de Biologie, 1015 Lausanne, Switzerland
Roger G. Symes Wildlife and Storage Biology Department, Agricultural Development and Advisory Service, Burghill Road, Westbury-on-Trym, Bristol BS10 6NJ, UK
John A. Vaughan ADAS Worplesdon Laboratory, Ministry of Agriculture, Fisheries and Food, Tangley Place, Worplesdon, Guildford GU3 3LQ, UK
David Welch Institute of Terrestrial Ecology, Hill of Brathens, Banchory, Kincardineshire AB3 4BY, UK
John Whelan Department of Agricultural Zoology and Genetics, University College, Belfield, Dublin 4, Ireland

One

Introduction: mammals as pests

R.J. PUTMAN

WHAT IS A PEST?

The Oxford English Dictionary defines a pest as: 'Any thing or person that is noxious, destructive or troublesome'. A variety of other definitions exist in the biological literature, as for example: 'a living organism which causes damage or illness to Man or his possessions or is otherwise, in some sense, "unwanted"' (Conway, 1981), but most biological definitions include some consideration of the economic significance of the damage caused. Thus 'A pest is an organism which harms Man or his property or is likely to do so. The harm must be significant, the damage of economic importance' (Woods, 1974). This last distinction is I feel an important one: much time and effort has been devoted in the past to the control of animal populations whose activities, while doubtless of considerable nuisance value were perhaps, if the situation were viewed more objectively, of no real economic significance. In such situations costs of control quite frequently exceed the real costs of any damage caused. The papers and articles in this book are edited and revised selections from presentations made at a Symposium on 'Mammals as Pests' hosted by the Mammal Society in November 1987. In organizing that Symposium I stressed the importance of considerations of economic significance and defined a pest as: any animal species or population which by its activities conflicts with Man's interests to a level where the damage caused becomes of economic significance. Indeed, one of the main purposes of this book is to re-evaluate the true pest status – in terms of real economic damage caused – of many common species traditionally regarded as pests.

Even such a definition is fraught with difficulty, however, for to some extent, pestiness is in the eye of the beholder. Through the end of the nineteenth century and up until the introduction of myxomatosis in the 1950s there were estimated to be some 100 million rabbits in Britain, and damage caused to agricultural crops was catastrophic. Rabbit numbers are once again increasing as myxomatosis weakens its control of populations. The estimated cost of rabbit damage by 1984 was between £95m and £120m annually with the potential to rise to between £240m and £400m (Rees *et al.*, 1985). Under a definition requiring significant economic damage rabbits are indubitably a pest (arguably the most significant of all agricultural pests), and they are indeed considered as such within these pages. However, some would argue on conservational grounds that the return of the rabbit should be welcomed. It has long been accepted that the characteristic rich flora of chalk downlands, developed under the heavy grazing pressure of the vast sheep flocks of the mediaeval graziers, has since been maintained by rabbits – since rabbit grazing, like that of the earlier sheep flocks, will arrest the plant succession, prevent the establishment and growth of scrub, and inhibit the dominance of aggressive grass species which might outcompete the less vigorous herbaceous species so characteristic of the downland sward (Tansley, 1922; Tansley and Adamson, 1925; Hope-Simpson, 1940). Hence we might argue that, much as we may define a weed as a plant in the wrong place, or a pollutant as a perfectly respectable chemical in the wrong context or in inappropriate concentration, so some animal pests too are only pests when in inappropriate numbers or in the wrong context.

Clearly this is not true in all cases but it is certainly true that in many instances pest problems are characteristic of populations of a particular species which have for some reason or another reached excessively high densities, or occur in unnatural contexts, so that their pest status results from an imbalance in the natural order. Indeed, in this country and elsewhere many pest problems are associated with non-native species: species of animals (or plants) introduced into new contexts often in the absence of their natural predators, where thus they have had no opportunity for long-term coevolution with other parts of the ecosystem – the plant or animal species on which they feed or the predators or competitors which might control their abundance. It can be no coincidence that of the species considered in these pages, brown rats, grey squirrels, coypu, mink, rabbits, sika and muntjac deer are all introduced species to Europe; the impact of (introduced) rabbits on the Australian economy or of the (introduced) possums or red deer, fallow, sika and white-tailed deer on forest management in New Zealand is well known (see for example Challies, 1985).

Even when the pest species may not itself be introduced, a lack of natural balance with the surrounding ecosystem may equally result if Man's interference has altered the environment to such a degree that, again, the animal no longer finds itself within the system to which it was originally adapted. Thus excessive habitat erosion, with loss of natural habitats to agriculture or industry may force an animal into unnatural contexts; progressive reduction of diversity in agriculture and other managed ecosystems may again threaten the balance of a co-adapted system, and finally, Man's production for his own ends of artificial communities (agricultural fields, timber plantations, grain stores and feedlots) may in itself create artificial but 'super-optimal' environments for the increase of popular species. Such drastic habitat change perhaps underlies the current pest status of roe deer, foxes and commensal rodents.

Such a synthesis is obviously oversimplistic and it is clear that many other species come into conflict with Man's interests merely in the context of their normal ecology. Other features tend to characterize pests however. The majority are 'r-strategists' (MacArthur and Wilson, 1967 and for example Horn, 1978): opportunistic species characteristic in general of early successional, productive environments; species which are ecologically flexible and can adapt readily to a wide range of circumstances, taking advantage of new opportunities, new environments. All are highly fecund and have high dispersal rates (factors which assume considerable importance in attempts at control through reduction of population numbers: pp. 6–7 and see Chapters 2, 9, 12, 14).

In this volume a pest is defined as 'any animal species or population which by its activities conflicts with Man's interests to a level where the damage caused becomes of economic significance'. Such creatures may conflict with Man's interests as competitors – eating foods or other products destined for consumption by Man or his domestic livestock – or as predators: preying upon animals maintained by Man in domestication. Damage may be direct or indirect; in many instances the economic significance of damage caused relates not so much to the initial damage itself but to the fact that secondary infections may enter a crop through wounds inflicted; it is commonly this secondary, bacterial or fungal infection which causes the more serious economic loss (e.g. Hampson, 1984; Crawley, this volume). Finally, in these pages we also consider animals which may act as vectors for diseases of economic significance. Badgers have been implicated in the spread of bovine tuberculosis (Zuckerman, 1980; Anderson and Trewhella, 1985), commensal rodents may be a serious source of a variety of zoonotic diseases (Chapter 5) and foxes are a serious potential vector of the rabies virus (Chapter 20).

The behaviour of any animal that brings it into conflict with Man may be part of its normal behavioural repertoire or may result from abnormal behaviour. Thus damage by rodents to grain stores or by rabbits to growing crops is occasioned during normal foraging activity; fraying of young saplings by roe deer may equally result during the perfectly normal behaviour of cleaning the protective 'velvet' from growing antlers or by the thrashing of vegetation with the antlers during aggressive display. By contrast, bark-stripping of trees by grey squirrels does not occur in their native America (Chapter 13) and while thrashing of vegetation by roe bucks during aggressive display may occur as a perfectly normal behaviour the frequency of incidence may be increased in areas where the natural population-structure has been distorted (e.g. Putman, 1985).

PEST MANAGEMENT

Traditionally, the approach to control of a pest problem, worldwide, has been attempted reduction of the population of that species considered to be causing damage. In many situations this may indeed be the appropriate response. However, it has been becoming increasingly clear over the years that there are many problems with such an approach.

By the very nature of the opportunistic boom and bust reproductive strategy that characterizes pest species, any population reduction achieved, short of extermination, is bound to be short-lived. The response of any wild animal population to a local reduction in density is to increase recruitment: through increased reproductive rate or increased juvenile survival within the population itself, or through an increase in immigration of individuals from nearby populations. Keeping populations of potential pest species below a level at which they may cause economic damage thus requires a continued input of effort. Indeed – just as in the worst forms of drug addiction – continued control of pest populations may require ever-increasing efforts merely to achieve the same effect.

If such management policies are poorly conceived they may even result in an overall *increase* in population number, disruption of population structure (with consequent disruption of social behaviour), fragmentation of the population and dispersal to areas hitherto free of damage: all of which may in effect result in an overall increase in economic levels of damage sustained (see for example review in Putman, 1985). Culling by rifle or shotgun can be selective; traps and poisons are rarely so. Further, continued and regular use of poisons may result in the development within the target population of strains resistant to such toxins. Yet, rather ironically, objective consideration of many situations where direct pest contol of this kind has been imposed in the past suggests that in a number of cases such control was in fact unnecessary or uneconomic from the outset. In many cases the attributed damage was not in fact of any economic significance in the first place or the losses sustained less than the cost of attempting control.

Such considerations are not unique to vertebrates. The very same problems (of the continuous and increasing efforts required to achieve reduction of populations below economic thresholds, of the rapid development of resistance to pesticidal chemicals) have been appreciated in the field of insect pest control for many years. Indeed, the parallels are striking. Yet, while in the field of invertebrate pest control development of new technologies and expertise has proceeded apace, response in management of vertebrate pests has been slow.

Control of insect pests now generally involves some sort of integrated management involving both control of the pest population itself but also control of its pest status (e.g. Conway, 1981; Horn, 1988). Some degree of chemical control of population size is exercised but this is frequently combined with efforts at extrinsic or intrinsic biological control to maintain populations at the new reduced level – in the use of natural enemies or in disruption of breeding patterns through sterile release or pheromonal control (e.g. Price-Jones and Solomon, 1974; Samways, 1981). At the same time, reduction of the economic significance of damage caused has been approached through the development of tolerant or resistant strains of crop plants.

The same principles, or ideas, could with advantage be applied to control of vertebrate pest problems (see for example Hansson and Nilsson, 1975). Control by predators may not be a realistic option in many cases, since vertebrate predators rarely have the potential to regulate their prey. However, certain other forms of biological control by natural enemies (parasites, diseases) may well warrant investigation. Myxomatosis was, at least in the short term, remarkably effective in achieving control of rabbit populations in the 1950s and 60s (Ross and Tittensor, 1986). Control of population size by pheromonal disruption of breeding or by physical or chemical sterilization is also practicable in certain situations (Jackson, 1972; Rowe and Lazarus, 1974a, b; Chapter 6 this volume; Matschke, 1977, 1980; Neville, 1982, Chapter 24 this volume), while perhaps the area with greatest potential is in manipulation of the actual pest status of the mammal concerned. If by careful research we can establish why an animal performs a particular behaviour pattern which causes it to conflict with Man's interests, we may be able to alter the environment in some appropriate way so that it performs that behaviour elsewhere, or no longer needs to perform it. Alternatively we may be able to modify the behaviour directly, or even exploit our knowledge of the animal's whole behaviour repertoire in order to minimize its pest status. Such methods might offer a greater potential for real control and perhaps a more permanent solution to the conflict. One final and important lesson vertebrate managers might learn from insect pest control is that the most effective management exercises overall are those which integrate a number of separate approaches to the problem.

It seems appropriate to summarize these arguments here in more detail. My comments are not restricted to any single species; indeed, I hope to develop through this paper a general framework applicable to any vertebrate pest, for objective evaluation of all possible management alternatives. But although the 'ethos' I present is applicable to all species, I shall illustrate it largely by reference to deer control – a bias justified by personal prejudice, and by the fact that they are the group I know most about (see also Putman, 1985).

To begin with then, I will attempt briefly to review the pros and cons of simple population control as a method of containing a vertebrate pest problem. The advantages are clear-cut. It is an obvious solution, and in the short term may prove effective. The disadvantages are less obvious and tend towards the insidious. First, as already noted, any reduction in population number achieved is likely to be short-lived. Total extermination of any vertebrate pest is well-nigh impossible, except in the most unusual circumstances (perhaps a population enclosed in an effective ecological island); the literature is cluttered with accounts of the failures. More commonly, managers are satisfied with a short-term reduction in numbers, but as noted this means that any control policy based on population reduction must be sustained and repeated year after year; both financially and in terms of manpower a costly decision. Quite simply, many populations which have reached a density at which they constitute a problem as a pest are probably already suffering reduction in recruitment and survival due to the effects of that same density. If population levels are lowered artificially, the density-dependent brake on population growth is released, reproduction increases, mortality declines. Such response to 'predation' in animal populations is widely recognized (but in practice there exists surprisingly little empirical documentation of the effect for wild vertebrates: Harris and Kochel, 1981). Even in populations where we cannot prove such density-dependent effects on recruitment and survival, local reduction of the species in one area is rapidly compensated for by immigration from outside. Population reduction is thus a perpetual treadmill. If carefully carried out it may offer short-term (and quick-acting) relief of pest pressure in particularly badly affected areas. As a long-term solution it is an extremely expensive proposition to contemplate.

My remarks so far have assumed a population reduction intelligently planned. Indiscriminate culling, or a cull concentrating on the wrong age or sex class of animal, can not only prove expensive but counter-productive. We have already noted that the disturbance caused by trapping or shooting may cause fragmentation and dispersal of the population to new areas, or may cause a shyness of behaviour which makes future control even more difficult. A clear illustration of this last point may be taken from consideration of the red deer populations in New Zealand. The introduction of red deer to New Zealand and their subsequent rapid increase to pest proportions is well documented (Challies, 1985). Government-promoted control schemes were introduced in 1931 and have continued in various forms to the present day. Control operations initially involved hunting on foot, with or without dogs. Although high cull figures were achieved, numbers of deer killed per man remained constant over many years, suggesting that hunting had little real impact on deer numbers. The introduction of helicopter hunting improved access to more inaccessible terrain, and with the establishment of a new deer farming industry in the early 1970s, hunting intensified as the demand for live animals suddenly made such hunting extremely lucrative. Helicopter crews could catch or kill five or six deer per flying hour (up to 12 000 deer per helicopter year). The impact was colossal, but restricted largely to open areas on hill tops or screes. Numbers appeared to decline dramatically, to a point where in 1982, many helicopter crews went out of business because catch rates no longer justified the expense of the operation (Challies, 1985). Although there can be no doubt that numbers of red deer were indeed greatly reduced, it is unlikely that this alone contributed to the sudden decline in catch rate. Pellet group counts on open range suggest deer densities only slightly lower than those of the middle 1970s and it has been proposed that the real change is behavioural, that animals now positively avoid the habitats where they are more exposed to attack (Challies, 1985). If this is true, further culling of these feral populations becomes increasingly problematical.

Problems of control may be exacerbated where regular use is made of poisons as a major element of the control programme, through the development of resistance. Resistance may take two major forms: acquired phenotypic resistance by an individual from continuous exposure to sublethal dosages of a toxin, or genetic resistance in the evolution of resistant strains within a population. The degree of acquired resistance can be controlled by eliminating unnecessary poison baiting or over-use of poisons, and by ensuring that when poisoning is carried out dose rates applied are lethal. The problems of genetic resistance are harder to overcome: the spread of warfarin-resistance in rats across the UK is now well known (Greaves and Rennison, 1973; Greaves *et al.*, 1977). (In the same context, poisoning as a form of population control may in addition be undesirable due to unpredictable effects of acute toxins on non-target organisms). Similar problems of resistance may also arise in relation to the use of diseases for biological control. Resistance to myxomatosis in rabbits is due to the evolution of genetically resistant strains, and also to the individual immunity acquired by a rabbit recovering from infection by a less virulent strain of the myxoma virus (Ross and Sanders, 1977, 1984).

All this is not intended to suggest that control of a pest problem by direct reduction of population size is never appropriate. In many situations it may be the only realistic option available, and if carried out intelligently and with full understanding of the underlying dynamics of the pest population concerned, it may prove an effective method of control (e.g. Chapters 6, 9, 14). However, there are often alternatives to be explored. Perhaps the first, and the easiest alternative is *laissez-faire*. In many cases an alleged pest problem may prove on closer examination to be insignificant – or at least in economic terms, less costly to accept than to prevent. Adoption of even effective population control measures may prove more expensive than the loss sustained; risks of increasing damage through ill-considered or ineffective control measures may outweigh the costs of accepting current damage levels. Studies of the damage caused to cereal crops by roe deer in agricultural areas of Southern England (Putman, 1985, 1986) revealed that damage caused was highly seasonal, and that while roe grazed in arable crops from November until the end of May, causing extensive damage to the crop through trampling and grazing, they left the fields by mid-May. Cereals are capable of considerable compensatory growth so that by harvest in late July the actual economic significance of the *apparent* damage by roe was insignificant (Putman, 1986, but see this volume, Chapter 15). The number of grain-bearing stems in areas of barley and wheat fields grazed by roe until mid-May was not significantly reduced over that of controlled, ungrazed plots in the same fields. Ears reached maturity at the same height and by the same date as those of undamaged plants; overall grain yields were not significantly reduced (Putman, 1986).

Timing of damage in relation to the growth stage and growth characteristics of the crop will clearly affect the economic significance of any damage caused since it will markedly influence the degree of crop recovery possible after damage ceases. In a series of experimental trials where winter and spring sown plots of wheat and barley were machine-mowed on a regular basis to simulate deer grazing, significant reductions in the grain yield at harvest were recorded where mowing continued until the end of May or into June for both cereal types (Putman unpublished data, and Table 1.1), and some losses were sustained by wheat crops under any mowing regime. However, in the instances cited of roe deer grazing cereal crops in the wild, despite the dramatic appearance of the damage when first sustained, actual economic loss was insignificant. The pest problem was illusory and any costs of management would have been wasted.

Nor is this an isolated instance. It has been shown that loss of yield in sugar beet and maize crops in Czechoslovakia due to grazing of vegetative parts of the plants by roe deer was also likely to be insignificant (Obrtel and Holisova, 1983; Obrtel, Holisova and Kozena, 1984). Thus loss of vegetative parts of maize crops during the summer months resulted in a decrease in fresh weight of ears at harvest of only 2.6%. (Since, in addition, damage affected less than 0.7% of the crop, the effective loss of yield for the crop as a whole was less than 0.15%; Obrtel and Holisova, 1983). Kaluzinski (1982) calculated that despite high densities of deer in agricultural areas in western Poland, consumption of vegetative parts of cereal crops outside the growing season was <1% and would not significantly influence yield, and that although later damage involving direct removal of ripening grains caused measureable and permanent loss, it was none the less an insignificant proportion of the crop as a whole. Even where in this case actual economic loss is real, it may still be less costly than the expense of control measures required to reduce that loss.

These are of course especially selected examples, and often some form of management intervention is required and justified. Thus for example we have already noted that agricultural damage by rabbits in the UK may approach £240m–£400m (Rees *et al.*, 1985). Significant damage to forestry and agriculture by moose in Norway has been recorded by Hjeljord (1979). Damage by sika deer to forestry in Ireland has been assessed by Larner (1977), whereas in the UK itself damage caused by red and roe deer to Scottish conifer forests has been considered by Welch *et al.*, (1983) and Staines (1985). Deer have also commonly been implicated as serious pests of horticulture (Scott and Townsend, 1985) and orchard crops (review in Matschke, de Calesta and Harder 1984). Amongst the most serious mammalian pests worldwide are various species of rodents, and here again levels of damage caused to cereals and other agricultural crops can be demonstrated to be extremely serious (e.g. Wood, 1984; Hampson, 1984 and see review by Dubock, 1984). But my point is that in many cases some form of control is attempted, often at

Table 1.1 Effects of simulated deer grazing on winter and spring sown plots of wheat and barley (Putman, unpublished). All plots were presprayed with glyphosate (Roundup (c) Monsanto Ltd) and ploughed before sowing. Seed was broadcast and rolled in at a sowing rate of 140 kg per hectare, and plots were given a single application of top-dressing (N:P:K, ICI no. 3) just after full emergence of the seedling crop. All treatments were replicated in 6, 4 m × 4 m blocks. Winter sown crops (sown mid-October) were mown fortnightly from the time of full shoot emergence (mid-January) until the end of February, March, April, May or June. In each case one set of replicates remained unmown throughout. Fortnightly mowing of crops sown in Spring (March) commenced in mid-April and continued until the end of May or end of June. Again, one set of plots remained unmown as a control. All mowing was with a tractor-mounted rotary mower adjusted to a height of c. 40 mm above ground level. Plots were harvested by hand at the end of the second week of August. Results, presented here for the years 1986/7 only are presented as mean number of ears or grain-bearing stems per m^2, total dry weight of (trimmed) ears per m^2 (again as a mean over all replicates) and weight yield as a percentage of the value for control plots. Values significantly different from controls are asterisked at $P < 0.05$* and $P < 0.01$**. (Bracketed values are standard distinctions.)

Crop type	Yield	Treatment					
		Unmown	Mown till end February	Mown till end March	Mown till end April	Mown till end May	Mown till end June
Winter barley	No. of ears m^{-2}	205.2	208.8	188.0	312.8**	186.0	52.0**
		(71.2)	(30.4)	(30.0)	(59.6)	(46.0)	(17.6)
	Weight of ears m^{-2}	280.0	290.8	266.8	374.4*	107.6**	10.3**
		(100.8)	(40.4)	(70.4)	(16.4)	(18.0)	(4.6)
	Yield as % control	—	104%	95%	134%	38%	4%
Winter wheat	No. of ears m^{-2}	298.8	227.2**	354.0	363.2*	131.2*	123.2**
		(46.4)	(36.0)	(123.6)	(68.0)	(36.8)	(24.8)
	Weight of ears m^{-2}	620.4	411.6	440.0*	464.9**	27.3**	64.8**
		(130.8)	(88.8)	(140.0)	(47.2)	(4.0)	(18.0)
	Yield as % control	—	66%	71%	75%	4%	10%
Spring barley	No. of ears m^{-2}	776.0	—	—	—	500.8**	Nil
		(82.4)				(98.0)	
	Weight of ears m^{-2}	516.8	—	—	—	189.9	Nil
		(48.6)				(38.4)	
	Yield as % control	—				37%	
Spring wheat	No. of ears m^{-2}	498.4	—	—	—	176.0**	Nil
		(71.6)				(70.0)	
	Weight of ears m^{-2}	542.4	—	—	—	90.4**	Nil
		(77.6)				(43.2)	
	Yield as % control	—				17%	

(Seed varieties used were winter wheat: Brigand (PBI Ltd), winter barley: Pirate (Twyford Seeds Ltd, Oxford), spring wheat: Musket, spring barley: Apex (both Nickerson RPB Ltd) considerable expense, without such evaluation: without first stopping to consider whether or not such control is indeed indicated, or would be cost-effective.

Even where some form of intervention is required, population reduction by direct culling may not necessarily be the most obvious or cost-effective solution. Such a measure carries within it, as we have noted, tremendous risks of worsening rather than ameliorating the problem and even if effective can at best offer only short-term relief. One of the most effective mechanisms of biological control exploited in insect pest management programmes is in interfering with natural patterns of reproduction through sterilization, sterile release or pheromonal disruption. The potential of sterilization or pheromonal disruption have yet to be fully explored among mammals and may prove perhaps to be limited to particular rather specific sets of circumstances. However, pheromonal interference with rabbit reproduction is currently under investigation (Bell, 1987). Neville (1982; and Chapter 24, this volume) has already had considerable success in controlling the pest status of populations of feral cats by trapping and neutering them before returning them to site, and the potential for chemical sterilization of vertebrate pests with field baits has been considered for cats (Remfry, 1978), deer (Matschke, 1977, 1980), rats (Rowe and Lazarus 1974a, b) and squirrels (Johnson and Tait, 1983). Excellent reviews of the action of the different chemicals available and their potential for 'field' application may be found in Jackson (1972), Neville (1982) and Johnson and Tait (1983). Chemical sterilization offers the attraction that it may be possible – if using non-hormonal chemicals – to cause sterility while not affecting behaviour. Dominant individuals in social species and territory-holders in territorial species still remain dominant, or retain their territory, preventing outsiders from entering the population (and thus, behaviourally reducing immigration), while at the same time reproductive recruitment is reduced amongst the 'home' population by infertility. Although a variety of active chemicals have been identified in laboratory tests, it has proved difficult to apply this method to wild populations and the few field trials that have been attempted have not been particularly successful (Rowe and Lazarus, 1974a, b; Remfry, 1978; Lazarus, this volume). The system has numerous attractions, however, in using the behavioural characteristics of the animals themselves in regulating population numbers.

Such regulation is likely to achieve longer-term control of population size. Long-term reduction of pest status may also prove possible through manipulation, not of reproductive behaviour, but of the behaviour which brings a given pest species into conflict with Man. To my mind such control of a pest problem is the most elegant and intellectually satisfying of all possible control measures while potentially of considerable long-term effectiveness.

I have already noted (p. 8) that the use of arable crops by roe deer in Hampshire is highly seasonal. Roe deer are well known to be 'concentrate selectors' (Hofmann, Geiger and König, 1976) needing to feed selectively on small morsels of highly nutritious foodstuffs. Close examination reveals that the use of agricultural pastures and cereals corresponds very closely to a time when they are the only vegetational communities showing new growth (and thus offering maximum digestible nutrients). As soon as the grasses of woodland glades or rides begin to flush, or buds burst on woody browse species, the animals switch their attention to these preferred foodstuffs and stop feeding on crops. Growth of winter cereals or fodder crops such as turnips or kale in the autumn and winter likewise coincides with the end of the growth period in the more 'natural communities'. Should we feel a need to manage roe deer away from arable crops therefore, we might achieve this most successfully by some form of habitat alteration which would offer highly nutritious forage elsewhere in the animal's range at these critical times of year. If offered in equivalent quality but without the drawback of the necessary exposure experienced by deer feeding out in open fields, such 'honeypots' might well be preferred, and use of farmland would be correspondingly reduced.

Grassed clearings in woodland, artificially sown and artificially fertilized, were once an important feature of deer management throughout Europe – drawing deer to central 'lawns' where they would be counted more easily, and more easily shot – helping to keep them off open ground surrounding the deer forest. The establishment of an understorey in commercial woodland, of bramble and rose, hawthorn, blackthorn, willow and hazel, might also reduce the need for foraging outside the woodland, or on the commercial timber crop (while at the same time permitting an actual increase in deer density). Such diversionary feeding or habitat management has also been advocated by other authorities. It is only fair to note, however, that such management by manipulation is not always so effective and the evidence in the literature is conflicting. Even in this same context of prevention of deer damage to forestry, reports are contradictory. Stenin (1970) for example offers evidence that feeding of red deer reduces damage by browsing or bark stripping to forest timber; Fischer (1975) claims that provision of sugar blocks eliminated bark-stripping behaviour by the same species – and bark-stripping recommenced when the sugar blocks were withdrawn. By contrast, Ueckermann (1983) investigated the effects of a variety of different feed additives on the extent of bark-stripping by red deer and concluded that none of them had any effect whatsoever in reducing damage. Clearly more work needs to be done, but I firmly believe there is a real potential in this area and as noted, such behavioural manipulation, if effective, offers a potential for a permanent solution to a given pest problem.

Perhaps none of these methods can prove totally effective in isolation and we may see for control of vertebrate pests, as for invertebrates, a movement towards a combined programme of integrated control. Yet all these separate measures are worthy of consideration and the potential risks of ill-conceived and ineffective attempts of straight population reduction are certainly sufficient to encourage us to look elsewhere.

In 1984, I summarized these ideas in a presentation to a symposium organized by the Universities Federation for Animal Welfare on 'The Humane Control of Land Mammals and Birds' (UFAW, 1985, see also Putman, 1988). My paper came at the end of two days of presentations. I knew that none of my ideas were original, but I had not realized how widely they were now shared. At the end of that two-day meeting I had heard the same ideas rehearsed and reaffirmed in a dozen earlier papers: a remarkable and a heartening experience (even if somewhat daunting in light of the fact that I still had my own paper to present!). Listening to these same ideas being presented and endorsed by so many different people representing the widest possible range of interest (academic biologists, ministry officials, welfarists and pest-control operatives alike) made it clear to me that attitudes were changing, and changing fast.

It was the success of that meeting and the 'smell of changes in the wind' that led me to suggest to the Mammal Society, some three years later, that they should host a symposium on 'Mammals as Pests' to explore further this change in attitude and the new biological insights required to provide the input on which such a new approach to management would be based. The meeting would concentrate on mammals (but in doing so could expand the range of species to which consideration was offered beyond that covered in the earlier UFAW volume). In soliciting contributions from various authors I defined two clear but interrelated themes: asking for papers which re-examined the true pest status of allegedly pest species by reviewing objective data on the actual economic significance of damage caused or which reviewed new advances in our understanding of the biology of some established pest species (behavioural, ecological or physiological) which could be exploited in improving management policy. I deliberately focused on these two themes since I felt these highlighted areas where lack of objective evidence in the past had constrained imaginative management and that it was in these two areas where the greatest advances could now be made. Many of the papers in the book of course address both themes and authors were extremely patient with me in sticking to these topics and in relating their contributions to such context.

THE PRESENT VOLUME

In my presentation to the UFAW symposium in 1984, I concluded

> When faced with any pest problem, the responsible manager should set himself the following questions:
> 1. Whatever the apparent damage, is the animal pest causing actual significant economic loss?
> 2. Are the costs of any form of management likely to be greater than the economic loss caused by accepting the damage without action?
> 3. Why is the pest animal behaving the way it is? Is it possible and economic to try and avoid damage by changing or refocusing the behaviour, through habitat manipulation or direct behavioural interference?
> 4. If population reduction appears the only possible, or the most economic solution, can this be achieved by behavioural or physiological interference rather than by culling?
>
> If, having addressed himself to these alternatives the manager still feels culling is appropriate, then the decision must be accepted. The culling itself should be humanely executed and intelligently planned with full regard to current knowledge on the population dynamics of the species concerned, the aim being to maintain a biologically sensible age and sex structure in the residual population after culling, and having regard to the likely biological responses of that population to the cull.

These same ideas are here developed and applied to a wide range of mammalian species. Throughout, the emphasis is on evaluating the true pest status of any species (in actual economic significance) and the biology of its interactions with Man which result in its conflict with human interest. While the papers in the book were of course contributed as separate articles, it is to be hoped that they may also be read together as a single, integrated whole.

Papers are grouped into specific sections, each relating to a particular species or related group of species. Each section begins with an introductory review of the pest status of the species concerned and an evaluation of their current economic significance. (In some cases these introductions take the form of separate chapters in their own right (Chapters 2, 7, 18, 22); in others they form the introduction to the first research paper in the section (Chapters 9, 12, 14, 15, 21). Each overview is followed by detailed papers selecting and concentrating on a particular, more restricted, topic and considering advances in our understanding of the biology of pest problems which may prove applicable to improvement of future control strategies.

The ideas rehearsed in the UFAW Symposium of 1984 and reiterated in this Introduction are reaffirmed throughout. Dunstone and Ireland (Chapter 21) and Anderson *et al.* (Chapter 23) conclude that mink and seals are perhaps not the pests they have for so long and so widely been assumed to be and that costs of attempted population control are thus perhaps unjustified. It is clear that although badgers may still pose a problem in the dispersal of bovine tuberculosis (a matter not as yet completely resolved: (Zuckerman, 1981; MAFF, 1981, 1982, 1983)), their economic significance as competitors or predators of man's livestock or agricultural crops, while possibly of local significance is not of national economic importance (Symes, Chapter 18; Roper, Chapter 19). Moles, too, must be considered primarily of nuisance value rather than as true economic pests (Stone, Chapter 7). By contrast we are left with little doubt that commensal and fossorial rodents can in many circumstances be significant pests (Richards, Chapter 2; Lazarus, Chapter 8) and deer and rabbits may equally clearly cause significant economic damage to silviculture and agricultural interests (Staines and Welch, Chapter 11; Crawley, Chapter 15).

Where pest status is substantiated the implications of proper consideration of population dynamics and social organization, in formulating effective management policy via direct population control, are explored for deer (Ratcliffe, Chapter 9), grey squirrels (Gurnell, Chapter 12) and rabbits (Cowan *et al.*, Chapter 16), while the efficacy of such control measures in containing the spread of a rabies infection through urban fox populations is examined by Smith and Harris in Chapter 20. The likely effectiveness of a variety of possible alternatives to direct population control are reviewed for commensal rodents by Lazarus (Chapter 8) and for feral cats by Neville (Chapter 24), while the potential of exploiting our knowledge of social behaviour in controlling pest densities or reducing the conflict with Man's interests is explored among moles (Gorman and Stone, Chapter 8), rabbits (Cowan *et al.*, Chapter 16), and coypu (Gosling and Baker, Chapter 14). Finally the potential of behavioural or ecological manipulations designed to repel pests or reduce their conflict with Man's interests is considered among rodents eating sugar beet seeds (Pelz, Chapter 3) and grey squirrels (Kenward, Chapter 13).

Throughout, all authors emphasize the importance of proper objective data in assessing the true scale of any pest problem, in predicting its likely repercussions – and in designing management plans calculated to contain it.

ACKNOWLEDGEMENTS

In writing this introduction I have drawn heavily (pp. 6–15) on material already presented in an earlier review at the UFAW Symposium on the Humane Control of Land Mammals and Birds in 1984. I would like to thank the organizers of that symposium for allowing me to quote so freely from that material, already published in the proceedings of their symposium (UFAW, 1985). In preparing the rest of the volume I am naturally deeply indebted to all the authors who succumbed to my blandishments and contributed the excellent articles brought together in this book. I thank them for agreeing to present articles at all, for directing their work so carefully within the context of my declared themes for the volume – and for putting up with my continuous editorial interference.

Many other names hide behind this book. Each of the papers included was carefully reviewed for me by a number of independent referees, whose comments were without exception most thorough and constructive. Refereeing manuscripts is a thankless task and I can only thank the many who so cheerfully and conscientiously agreed to shoulder the work. As editor, I certainly benefitted tremendously from their careful appraisal, I am sure the authors of each individual paper in the volume would also acknowledge how helpful the comments of these various referees have been to them in revising their draft texts for publication. While preserving the anonymity of referees for individual papers, I would like to express my thanks to Paul Chanin, Nigel Dunstone, John Gurnell, Tony Hardy, Steve Harris, Steve Havers, Colin Pringle, Phil Ratcliffe, Jonathan Reynolds, Chris Richards, Robert Smith and Brian Staines.

As noted earlier, the papers and articles in this book are selected from amongst those presented to a symposium organized in London by the Mammal Society in 1987. I would thus also like to thank all those who contributed to the success of that meeting: all speakers (including those not represented in these pages), Chairmen (Professor Sam Berry, Dr Tony Hardy and Dr Hans Kruuk) and the Mammal Society's Conference Secretary, Dr Stephen Harris, who shared with me the burden of arranging that symposium.

Finally I would like to thank the editor Bob Carling, sub-editor Simon Armstrong and everyone else at Chapman and Hall who guided this book so patiently through to publication.

REFERENCES

Anderson, R.M. and Trewhella, W. (1985) Population dynamics of the badger (*Meles meles*) and the epidemiology of bovine tuberculosis (*Microbacterium bovis*). *Phil. Trans. Roy. Soc. Series B*, **310**, 327–82.

Bell, D.J. (1987) Pheromones and mammalian pests, with particular reference to the European rabbit in *Control of Mammal Pests* (eds C.G.J. Richards and T.Y. Ku), Taylor and Francis, pp. 113–26.

Challies, C.N. (1985) Establishment, control and commercial exploitation of wild deer in New Zealand in *Biology of Deer Production* (eds P.F. Fennessy and K.R. Drew), Royal Society of New Zealand, pp. 23–36.

Conway, G. (1981) Man versus pests in *Theoretical Ecology: Principles and Applications* (ed. R.M. May) (2nd edn), Blackwell, Oxford, pp. 356–86.

Dubock, A.C. (1984) Economic benefits of vertebrate pest control in *The Organisation and Practice of Vertebrate Pest Control* (ed. A.C. Dubock), Imperial Chemical Industries, London, pp. 315–28.

Fischer, H. (1975) [Sugar licks prove successful]. *Die Pirsch – Der Deutsche Jager*, **3**.

Greaves, J.H., Redfern, R., Ayres, P.B. and Gill, J.E. (1977) Warfarin resistance: a balanced polymorphism in the Norway rat. *Genet. Res. Camb.*, **30**, 257–63.

Greaves, J.H. and Rennison, B.D. (1973) Population aspects of warfarin resistance in the brown rat, *Rattus norvegicus*. *Mammal Rev.*, **3**, 27–39.

Hampson, S.J. (1984) A review of rodent damage to sugar cane, with criteria for the use of rodenticides in *The Organisation and Practice of Vertebrate Pest Control* (ed. A.C. Dubock), Imperial Chemical Industries, London, pp. 227–52.

Hansson, L. and Nilsson, B. (1975) *Biocontrol of Rodents*. Ecology Bulletin *19*, Swedish National Science Research Council, NFR.

Harris, L.D. and Kochel, I.H. (1981). A decision-making framework for population management in *Dynamics of Large Mammal Populations* (eds C.W. Fowler and T.D. Smith), Wiley Interscience, Chichester, pp. 221–39.

Hjeljord, O. (1979) Damage to forestry and agriculture caused by moose in Norway in *Moose Management in Fennoscandia* (ed. S. Myrberget), Direktorat for vilt og ferskvannsfisk, Trondheim, pp. 34–40.

Hofmann, R.R., Geiger, G. and Konig, R. (1976) Vergleichend-anatomische untersuchungen an der vormagenschleimhaut von Rehwild (*Capreolus capreolus*) und Rotwild (*Cervus elaphus*). *Z. Saugetierkünde*, **41**, 167–93.

Hope-Simpson, J.F. (1940) Studies of the vegetation of the English Chalk VI. Late stages in succession leading to chalk grassland. *J. Ecol.*, **28**, 386–402.

Horn, D.J. (1988) *Ecological Approach to Pest Management*. Elsevier, Amsterdam.

Horn, H.S. (1978) Optimal tactics of life-history and reproduction in *Behavioural Ecology* (eds J.R. Krebs and N.B. Davies) Blackwell, Oxford, pp. 411–29.

Jackson, K. (1972) Chemical methods of male contraception in *Artificial Control of Reproduction* (eds C.R. Austin and R.V. Short), Cambridge University Press, Cambridge.

Johnson, E. and Tait, A.J. (1983) Prospects for the chemical control of reproduction in the grey squirrel. *Mammal Rev.*, **13**, 167–72.

Kaluzinski, J. (1982) Composition of the food of roe deer living in fields and the effects of their feeding on plant production. *Acta theriologica*, **27**, 457–70.

Larner, J.B. (1977) Sika deer damage to mature woodlands of south-western Ireland. *Proc. XIII Congr. Game Biol.*, pp. 192–202.

MAFF (1981, 1982, 1983) *Bovine tuberculosis in badgers*. Annual reports. HMSO, London.

MaCarthur, R.H. and Wilson, E.O. (1967) *The Theory of Island Biogeography*. Princeton University Press, Princeton.

Matschke, G.H. (1977) Anti-fertility action of two synthetic progestins in female white-tailed deer. *J. Wildl. Mgmt.*, **41**, 194–6.

Matschke, G.H. (1980) Efficacy of steroid implants in preventing pregnancy in white-tailed deer. *J. Wildl. Mgmt*, **44**, 756–8.

Matschke, G.H., de Calesta, D.S. and Harder, J.D. (1984) Crop damage and control in *White-tailed Deer: Ecology and Management* (ed. L.K. Halls), Stackpole Books, Pennsylvania, Ch. 38.

Neville, P.N. (1982). Gossypol: a literature review of Gossypol as an oral contraceptive and an assessment of its potential in the control of feral cats and dogs. Universities Federation for Animal Welfare.

Obrtel, R. and Holisova, V. (1983). Assessment of the damage done to a crop of maize (*Zea mays*) by roe deer (*Capreolus capreolus*). *Folia Zool. (Brno)*, **32**, 109–18.

Obrtel, R., Holisova, V. and Kozena, I. (1984) Deer damage to sugar beet leaves. *Folia Zool. (Brno)*, **33**, 99–108.

Price-Jones, D. and Solomon, M.E. (1974) *Biology in Pest and Disease Control. British Ecological Symposium*, **13**.

Putman, R.J. (1985) Alternatives to direct population control. (A review of the potential of behavioural manipulation in pest control) in UFAW: Humane Control of Land Mammals and Birds. Universities Federation for Animal Welfare, pp. 124–31.

Putman, R.J. (1986). Foraging by roe deer in agricultural areas and impact on arable crops. *J. Appl. Ecol.*, **23**, 91–9.

Putman, R.J. (1988) *The Natural History of Deer*. Christopher Helm, Bromley.

Rees, W.A., Ross, J., Cowan, D.P., Tittensor, A.M. and Trout, R.C. (1985) Humane control of rabbits, in UFAW: *Humane Control of Land Mammals and Birds*. Universities Federation for Animal Welfare, pp. 96–102.

Remfry, J. (1978) Control of feral cat populations by long-term administration of megastrol acetate. *Vet. Rec.*, **28**, 403–4.

Ross, J. and Sanders, M.F. (1977) Innate resistance to myxomatosis in wild rabbits in England. *J. Hyg.*, **79**, 411–5.

Ross, J. and Sanders, M.F. (1984) The development of genetic resistance to myxomatosis in wild rabbits in Britain. *J. Hyg.*, **92**, 255–61.

Ross, J. and Tittensor, A.M. (1986) Influence of myxomatosis in regulating rabbit numbers. *Mammal Rev.*, **16**, 163–8.

Rowe, F.P. and Lazarus, A.B. (1974a) Effects of an estrogenic steroid on reproduction of wild rats, *Rattus norvegicus*. *Agro-ecosystems*, **1**, 57–68.

Rowe, F.P. and Lazarus, A.B. (1974b) Reproductive activity in a wild rat, *Rattus norvegicus*, population treated with an estrogenic steroid. *Agro-ecosystems*, **1**, 227–35.

Samways, M.J. (1981) *Biological Control of Pests and Weeds*. Studies in Biology, **132**, Edward Arnold, London.

Scott, J.D. and Townsend, T.W. (1985) Characteristics of deer damage to commercial tree industries of Ohio. *Wildl. Soc. Bull.*, **13**, 135–43.

Staines, B.W. (1985) Humane control of deer in rural areas. in UFAW: *Humane Control of Land Mammals and Birds*. Universities Federation for Animal Welfare, pp. 105–10.

Stenin, G. (1970) [Feeding red deer and its effect in reducing damage]. *Gorsko Stopanstvo*, **26**, 40–4.

Tansley, A.R. (1922) Studies of the vegetation of the English Chalk II. Early stages of the redevelopment of woody vegetation in chalk grassland. *J. Ecol.*, **10**, 168–77.

Tansley, A.R. and Adamson, R.W. (1925) Studies of the vegetation of the English Chalk III. The chalk grasslands of the Hampshire–Sussex border. *J. Ecol.*, **13**, 177–223.

Ueckermann, E. (1983) [The effect of various feed-components on the extent of bark-stripping by red deer]. *Z. Jagdwiss*, **29**, 31–47.

UFAW (1985) *Humane Control of Land Mammals and Birds*. Universities Federation for Animal Welfare.

Welch, D., Staines, B.W., Scott, D. and Catt, D.C. (1983) Browsing damage done by red deer to young stands of sitka spruce. Institute of Terrestrial Ecology Research Reports (1983), 10–12. ITE, Cambridge.

Wood, B.J. (1984) Rat pests of tropical crops – a review of practical aspects in *The Organisation and Practice of Vertebrate Pest Control* (ed. A.C. Dubock), Imperial Chemical Industries, London, pp. 265–87.

Woods, A. (1974) *Pest Control*. McGraw-Hill, New York.

Zuckerman, O.M. (1980) *Badgers, Cattle and Tuberculosis*. HMSO, London.

Zuckerman, O.M. (1981) The great badger debate. *Nature Lond.*, **289**, 628–30.

Two

The pest status of rodents in the United Kingdom

C.G.J. RICHARDS

SUMMARY

A rodent population may be considered a pest if it causes damage which is substantially greater than the economic or environmental cost of controlling it. In urban areas, *Rattus norvegicus* and *Mus domesticus* are major public health pests. The average frequency of infestations of buildings by each of these species is modest (about 4% of premises) but 50% of premises or more may be infested in some areas. Little attention has been paid to whether or not this represents a satisfactory level of control. 'Rat-free' towns might be attainable by controlling *R. norvegicus* in refuge habitats. Improvement in *M. domesticus* control would require a change from a complaints procedure to area control. In rural areas, *R. Norvegicus* is a major pest around agricultural buildings. Existing control methods will be adequate in most areas for many years but behavioural and physiological changes within *R. norvegicus* populations in and around Hampshire need further study. Field rodent problems in the UK are very limited. Significant outbreaks of *Microtus agrestis* have not occurred for several decades, possibly as a result of changing farming practice. Other microtine rodents which are damaging elsewhere do not occur in the UK. *Apodemus sylvaticus* causes sporadic damage to sugar beet. Eradication of feral *Ondatra zibethicus* has been achieved and that of *Myocastor coypus* now seems attainable. The only rodent of economic significance in the field is *Sciurus carolinensis*, which debarks deciduous trees; here anticoagulants and trapping provide reasonable control.

INTRODUCTION

Fifteen species of rodents are established in Britain (Corbet and Southern, 1977) and at least twelve of these have been classified as pests at one time or another (Table 2.1). The definition of a pest, however, tends to be somewhat arbitrary; almost any species may cause damage or irritation in some situations, which may lead it to be branded as a pest. A more meaningful definition arises when the damage reaches a level at which some kind of action is demanded. With many agricultural pests, this is generally taken as the economic injury threshold. This is the level at which the economic value of damage is substantially greater than the cost of preventing it. It is implicit in this definition that the objective is to reduce the damage, not simply or even necessarily the pest. Thus the concept of controlling pest species is replaced with that of managing pest populations. Both sides of this equation, however, may be difficult to quantify. In the case of mammals, the definition is made more complex by value judgements associated with the candidate pest. It is not clear how, for instance, the aesthetic value of a species can be set against economic damage. The damage side of the equation is complicated in the case of public health, on which no economic value can be placed. A workable definition which will be used for rodents in this paper is: 'A pest is a population causing damage (or potentially causing damage) substantially greater than the economic or environmental cost of managing it'.

To set up such a definition, however, implies that damage should be quantified. Unfortunately, almost no quantitative data are available on the damage caused by rodents in the UK. This significant gap needs to be filled. Without at least semi-quantitative data, it is not possible to make rational management decisions in sensitive or borderline situations.

The difficulty of balancing rodent damage against aesthetic considerations is illustrated by the current legislation. The Prevention of Damage by Pests Act 1949 places a specific duty on local authorities to control 'rats and mice' and on the occupiers of buildings to report or control infestations. Commensal rodents are, therefore, scheduled as pests at any level of population size. The situation for field rodents is more complex. In this case, the law seeks mainly to define what is not permitted. The Protection of Animals Act 1911 prohibits the poisoning of

Table 2.1 The pest status of rodents in the UK

	Introduced	Habitat (distribution)	Damage
Muridae			
Apodemus sylvaticus		Fields and woodland	Sugar beet (sporadic)
Apodemus flavicollis		Ubiquitous (south)	None
Micromys minutus		Dense vegetation (south and central)	Historical in corn ricks
Mus domesticus	I	Commensal	Public health
Rattus rattus	I	Commensal (port areas)	Public health
Rattus norvegicus	I	Commensal/hedgerows	Stored crops, buildings, public health
Cricetidae (subfamily Microtinae)			
Clethrionomys glareolus		Fields and woodland	Historical only
Microtus agrestis		Rough grassland	Historical only
Microtus arvalis	I	Rough grassland (some islands)	None
Arvicola terrestris		By water	Negligible, very local
Gliridae			
Glis glis	I	Woodland (Chiltern Hills)	Debarking, nuisance
Muscardinus avellanarius		Dense woodland	None
Carpromyidae			
Myocastor coypus	I	Marshland (East Anglia)	Crops and dykes (now negligible)
Sciuridae			
Sciurus vulgaris		Coniferous woodland (patchy)	Debarking (limited)
Sciurus carolinensis	I	Woodland (patchy)	Debarking

any mammals, except 'rats, mice and other small ground vermin' where necessary to protect public health or agriculture. This Act does not attempt to define the relevant species but 'small ground vermin' has been assumed to include all small rodents but to exclude any mammal as large as a squirrel or rabbit. Thus it was considered necessary to issue the separate Warfarin Order 1968 to allow the poisoning of *Sciurus carolinensis*. The 1911 Act has been somewhat modified by the Wildlife and Countryside Act 1981, which protects wildlife in general. This specifically shelters *Sciurus vulgaris* and the Gliridae (i.e. both *Muscardinus avellanarius* and the introduced *Glis glis*), alongside other mammals such as badgers, bats and shrews. All small rodents other than the Gliridae presumably therefore remain 'small ground vermin' and can be killed. The scientific basis for these distinctions is not clear but the Act has the great advantage that protected species can only be controlled under licence from the relevant authorities. This does at least allow rational management decisions to be made for these species on a case by case basis. Interestingly, the 1981 Act is also taken to allow the poisoning of rabbits.

Different perspectives are nicely illustrated by *Rattus rattus*. This is scheduled as a pest under the 1949 Act but is now so rare that it has been considered for inclusion in the Red Data Book.

The scale of rodent pest damage in the UK, however defined, is indicated by the resources devoted to its management. This has been estimated at some £50 million p.a. (Meehan, 1984).

URBAN RODENTS

Distribution and management

Although most small rodents are recorded periodically in urban areas, the only ones regarded as pests are the commensal species *Rattus norvegicus*, *R. rattus* and *Mus domesticus*. These three species carry a wide range of diseases (Gratz, 1984). While the threat from epidemics such as bubonic plague and murine typhus has now receded in the UK, major rodent-borne diseases still occur in other parts of the world and could presumably return if rodent numbers became high enough. Even at the lower rodent densities now prevailing, a number of rodent-borne diseases are still significant problems, including salmonellosis and leptospirosis, and the commensal rodents justifiably retain high priority as public health pests. Other types of damage, such as comsumption and spoiling of food and damage to buildings are not significant in comparison with this threat.

The responsibility for the control of rodents as public health pests in the UK rests with the Environmental Health Departments of the local authorities. Up to 1979, the Ministry of Agriculture, Fisheries and Food (MAFF) provided advice and guidance to the local authorities on rodent control; this responsibility has now passed to the Department of the Environment. MAFF conducted regular surveys of commensal rodent infestations based on indices of infestation (MAFF, 1975; Rennison and Drummond, 1985). Although these indices reveal little about the levels of damage caused, they do show some interesting trends.

R. rattus, which was once widespread in the UK, is probably now entirely confined to the dock areas of three major ports, where continual reinfestation from ships is likely. *R. rattus* was the only commensal rat in the UK from its introduction around 1100 until the introduction of *R. norvegicus* during the eighteenth century (Corbet and Southern, 1977). Its decline since then is undoubtedly due largely to competition with the larger species. This was probably assisted by changes in house construction, especially the substitution of thatch by tiles, which favoured the ground-living over the climbing species. *R. rattus* was confined mainly to harbour areas by the early part of this century. It seems now to be confined to a few individuals in the ports of Bristol and London; the only known breeding population is on Lundy Island (Anderson, personal communication). The further decline over the last 50 years has been attributed to improved hygiene and control measures, particularly around the docks of smaller quays and fishing ports (Bentley, 1964). However, this continuing decline is in contrast to the experience on the mainland of Europe. For instance, *R. rattus* has expanded its distribution substantially in inland parts of Holland and Belgium in recent years (Bosman, personal communication). The reasons for this difference are unknown and the possibility of a resurgence in the UK cannot be ruled out.

Inspections of premises by local health officials between 1966 and 1979 showed that some 4% of premises were infested by *R. norvegicus* and a further 4% by *M. domesticus* at any one time (MAFF, 1975; Rennison and Drummond, 1985). Very few (0.8%) were infested by both species. *M. domesticus* was the more frequent species in large towns but *R. norvegicus* predominated in rural areas. This presumably reflects the influence of field *R. norvegicus* populations in continually re-infesting rural premises. There is some indication that the frequency of *M. domesticus* infestations increased between 1966 and 1973; a slight decline from 1976 to 1979 is more certain. Rennison and Drummond (1985) tentatively attributed the decline to improved hygiene. A more likely explanation may be the introduction of more effective mouse toxicants (calciferol in 1973 and difenacoum in 1975), some years after warfarin had ceased to give adequate control.

The average level of rodent infestation conceals very wide variation between areas. Meyer and Drummond (1980), for instance, found that nearly 50% of premises inspected before a trial were infested with *M. domesticus* in their London study area. Across Central London as a whole, 40% of premises were infested each year (MAFF, 1975). Meyer and Drummond (1980) showed that a substantial reduction in infestation level was possible by changing the management strategy. The municipality, which is responsible for rodent control, normally acted only in response to complaints from householders; switching to a regular baiting programme in heavily infested areas achieved a substantial improvement in control at minimal extra cost.

Existing levels of infestation are the result primarily of action taken by the occupiers of premises and by local authorities under the 1949 Act. Local authorities do keep certain types of premises, such as food shops and restaurants, under surveillance and use the legislation to enforce rodent control where necessary. Little attention has been paid, however, to the wider problem of pest rodents in urban ecosystems or to whether the existing level of infestation (about 8% of premises infested by rodents at any time) represents satisfactory control. Apart from the scattered premises kept under surveillance, management is essentially driven by a complaints procedure. Thus the prevailing level of rodent infestation in the urban ecosystem as a whole depends largely on the public perception of rodent damage. Since the primary motivation for urban rodent control is public health, this assumes that there is a general awareness that rodents carry diseases. This is probably a reasonable assumption in many cases; although the public motivation is often based on a deep cultural distaste for rodents, this may serve to set a lower limit of tolerance than that needed to prevent disease transfer. In other cases, however, such as in the area studied by Meyer and Drummond (1980), the result was not satisfactory. The weakness of relying on public complaints is that some sections of the human population may be more tolerant of rodents than others.

In the case of *R. norvegicus*, a higher level of control than 4% of premises is almost certainly possible, by changing to a more active management strategy but without necessarily moving to expensive baiting programmes. The concept of 'rat-free towns' has been popular in parts of mainland Europe for some years and may well be cost-effective (Drummond, Taylor and Bond, 1977; Drummond, 1985; Myllymäki, 1986a). This involves an active campaign to eradicate existing infestations in the town followed by substantial improvement in environmental hygiene. Permanent baiting and/or frequent inspection of refuge habitats and invasion routes such as rubbish dumps, river banks and sewers, minimize re-infestation. Sewer rat control is routine in the UK but other elements of the rat-free town concept are not widely used. Nor has any serious attempt been made to determine whether a rat-free town is a desirable objective. It may well become highly desirable if resistance to rodenticides ever becomes a serious problem.

A 'mouse-free town' by contrast, is not an attainable objective. Improvement in mouse control could be achieved in critical areas by changing to an area baiting programme (Meyer and Drummond, 1980). There is growing evidence, moreover, that intensive but partial mouse control over many years has selected for mice which are not only tolerant of rodenticides but have also changed their behaviour so that they have become difficult to control. For instance, some mice in the centres of Birmingham and London now seem to avoid a wide range of control agents, including traps, baits and even empty bait boxes and bait trays (Meehan, personal communication). Clearly should this phenomenon become widespread, it would have serious consequences for mouse control. It also illustrates a long-term weakness in relying solely on a complaints procedure.

COMMENSAL RODENTS IN AGRICULTURE

Distribution and damage

There are few farms in the UK which are free of *R. norvegicus*. The farm ecosystem may usefully be divided into two compartments for *R. norvegicus* – the buildings and the fields. Although the species does not seem to be well adapted to field conditions, low density populations probably exist in hedgerows and shelter-belts throughout the year (Huson and Rennison, 1981). The carrying capacity of the fields may be higher in summer, when the insect and cereal foods on which *R. norvegicus* rely are more abundant. The availability of cereals may be critical to carrying capacity; there is certainly a correlation between the frequency of *R. norvegicus* infestations around farm buildings and the acreage of arable crops on the farm (Greaves, 1988). These field populations are not thought to cause significant damage to crops; only under exceptional circumstances are *R. norvegicus* sufficiently numerous to cause visible damage.

There is a strong seasonal variation in infestations around farm buildings (Huson and Rennison, 1981). This is due, in part, to seasonal movement of *R. norvegicus* between the field and building compartments of the farm ecosystem. The distance involved in such movements would certainly present no problems to *R. norvegicus* (Hardy and Taylor, 1979). After the cereal harvest, the carrying capacity of the fields is reduced, whereas that of the farm buildings increases as livestock are brought in for the winter and harvested grain is stored. The onset of colder winter weather then causes a sharp rise in farm building infestations. The carrying capacity of the buildings may reduce again in the spring, leading to a reverse movement. Even without this seasonal movement, emigration from uncontrolled infestations around buildings will certainly increase or re-establish field populations.

M. domesticus is less frequent than *R. norvegicus* around farm buildings (Rennison and Drummond, 1985). They are infrequently found in the field and this must help to break the cycle of eradication and re-infestation seen with *R. norvegicus*.

Once again there are few data on the amount of damage caused by commensal rodent infestations around farm buildings. While stored crops and animal feed are certainly consumed and spoiled, it is difficult to show that this is significant compared with general wastage. Perhaps the most significant form of economic damage stems from gnawing on electrical wires; some 50% of fires on farms have been attributed to this cause. All these sources of damage have been estimated at £10–20 million p.a. As with urban rodents, however, the primary motivation for control is generally human or animal health, with damage thresholds set low by cultural distaste.

Management

Since *R. norvegicus* populations around farm buildings are only part of wider farm populations, their management must consider also the populations in the field compartment of the ecosystem. There are, however, no obvious ways in which widely dispersed *R. norvegicus* in hedgerows could be managed. The use of rodenticides against such individuals is unlikely to be either cost-effective or acceptable in its effects on other animals. In practice, active management is confined to the vicinity of farm buildings. It may be possible to achieve a long-term reduction in field populations by managing *R. norvegicus* over wider areas, including a number of farms. Co-ordinated effort around buildings over the entire area in the winter may draw in individuals from the less favourable habitats of the field compartment. By reducing numbers in all buildings in the area, there may be fewer *R. norvegicus* in the fields the following year. Co-ordinated action over an extended period should break the cycle of infestation and re-infestation between buildings and fields. Such area control is certainly effective against rodents in analogous two-compartment ecosystems comprising rice-fields and houses in tropical villages (Richards and Buckle, 1986).

Management of rodents around farm buildings should rely largely on good hygiene. In practice, rodents are low priority on most farms and rodent infestations tend to be a symptom of general farm management style. Where farms are susceptible to rodent infestation, farmers tend to rely heavily on poison baiting.

There do not at present seem to be infestation problems of practical significance around UK farm buildings which cannot be adequately resolved by currently available techniques (Greaves, 1988). There is, however, one area of the UK which is receiving increased attention. In the area around Hampshire, extending into parts of Berkshire and Wiltshire, there are reports that *R. norvegicus* populations are becoming larger and more difficult to control (MacNicoll and Gill, 1987). Treatments with both anticoagulant and non-anticoagulant rodenticides take longer than in other parts of the country (Greaves, Shepherd and Quy, 1982; MAFF, 1982; Greaves, 1988). Laboratory studies have shown significant levels of resistance to some second-generation anticoagulants, especially difenacoum, in *R. norvegicus* from this area (Greaves and Cullen-Ayres, 1987). Although there is evidence for some increase in tolerance to bromadiolone and, to a lesser extent, brodifacoum, this does not seem to be enough in either case to be detectable in practical conditions and is not sufficient to account for the difficulties experienced in control operations. Any physiological resistance is greatly exacerbated by what appear to be unusual behaviour patterns in local *R. norvegicus* populations which result in them feeding less on any rodenticidal bait (Greaves, 1988). It is not clear whether the behavioural or the physiological phenomena are the more important but both aspects need further study. Are, for instance, the behavioural phenomena due to local ecological factors or are they some adaptation, perhaps genetic, to continual control efforts? This is critical for predicting the future spread of the problem. If the behaviour is not genetically based, the area affected may not increase; no geographical survey of the problem has been conducted but there does not seem to have been significant expansion on the east side of the area, which has been the most intensively studied (Greaves, personal communication).

With the exception of the Hampshire area, the outlook for farm rodent control in the UK is good. Warfarin resistance caused a major problem because it emerged rapidly and in several places. By 1972, some 20 years after the introduction of warfarin as a rodenticide, there were at least four different established warfarin resistance areas and resistant rats had been recorded in numerous other places (Greaves and Rennison, 1973). At least three different genetic mechanisms were known and the speed with which the resistance occurred suggests that the genes existed in the population before warfarin was used (Greaves, 1985). By contrast, cross-resistance to second-generation anticoagulants is limited. Only in one of the warfarin-resistance areas (Hampshire) are *R. norvegicus* now difficult to control. Resistance to difenacoum in Hampshire was, like warfarin resistance, noted very soon after its introduction in 1975, again implying that the genes existed before. This suggests that anticoagulant resistance either happens very quickly after the introduction of a new compound (i.e. the genes already exist) or only much more slowly (presumably once new genes appear). No areas of difenacoum tolerance have emerged outside Hampshire over 10 years of intensive use so it seems unlikely that resistance will become a problem elsewhere in the near future.

Moreover, *R. norvegicus* resistance areas spread rather slowly – about 4–7 km each year in the case of Welsh warfarin resistance (Greaves *et al.*, 1977). Even warfarin is still effective against *R. norvegicus* in many parts of the country. Finally, although the Hampshire *R. norvegicus* are difenacoum resistant, susceptibility to the more potent brodifacoum (Greaves and Cullen-Ayres, 1987) remains within the range expected for susceptible strains. A similar trend is evident in mainland Europe, where resistance to bromadiolone (which is approximately equitoxic to difenacoum for *R. norvegicus*; Dubock and Kaukeinen, 1978), has been reported from a number of countries. While warfarin resistance is widespread, resistance to bromadiolone and difenacoum is restricted to limited areas and no resistance to brodifacoum has been detected (Myllymäki, 1986b; Lund and Lodal, 1986). The Hampshire problem area may well spread slowly and the behavioural phenomena there require further study. While there is no room for complacency, existing methods will be used successfully for many years.

FIELD RODENTS IN THE UK

The absence of major field rodent problems in the UK is in striking contrast to mainland Europe. The reaons seem to lie both in the absence of some species and possibly in aspects of agronomic practice.

Microtine rodents, especially those species which have 'cyclical' populations, cause extensive damage in many countries (Myllymäki, 1979). In the UK, 'outbreaks' of *Microtus agrestis* used to occur. Massive populations on pasture in Scotland in the 1880s so depleted the grass that sheep died of starvation (Elton, 1942). As recently as the 1940s, damage on some forestry plantations was severe. Since that time, however, damage by *M. agrestis* has been limited to occasional reports of minor damage to plantations in the hills of Scotland and Wales. There is evidence that *Microtus* problems have also decreased during this century in other parts of Western Europe but less so in Central or Eastern Europe; changes in agronomic practice may be responsible (Myllymäki, 1979). In the UK, population cycles probably do not occur in natural grassland, although they could occur in extensive rough grassland monoculture (essentially young forestry plantations) due to the absence there of vacant habitat patches into which juveniles can disperse (Richards, 1985).

Microtines also cause damage to orchards in both Europe and North America (Myllymäki, 1979). The absence of *M. agrestis* damage to orchards in the UK is probably due to intensive grass management. The UK is also fortunate in the absence of fossorial microtines such as *Pitymys* spp. or *Microtus pinetorum*, which cause orchard damage elsewhere and cannot be controlled by grass management.

The fossorial form of *Arvicola terrestris*, which causes substantial damage in many parts of Europe (Meylan, Chapter 4), is also absent from the UK. The amphibious form of this species is confined to the banks of streams and rivers, where it causes no significant damage.

Amongst the murids, *Apodemus sylvaticus* damage to sugar beet occurs in East Anglia. The pattern of damage in the UK is similar to that in West Germany (Green, 1979; Pelz, 1986, and Chapter 3). *A. sylvaticus* only damage sugar beet when natural food is scarce and the sporadic damage caused is probably best managed by providing an alternative food supply (Pelz, Chapter 3).

Introduced rodents cause the most concern in agriculture. Two species of large amphibious rodents – *Ondatra zibethicus* and *Myocastor coypus* also provide rare examples of successful long-term management. Feral *O. zibethicus* escaped from fur farms in the 1930s but were eradicated by 1937. Intensive trapping of *M. coypus* in East Anglia, also escapees from fur farms, has now reduced the population to very low levels and the objective of eradication is in sight (Gosling and Baker, Chapter 14).

The only significant field rodent problem in the UK is squirrel damage to forestry. This is mainly caused by the introduced *Sciurus carolinensis*, which removes bark, especially in beech plantations (Kenward, Chapter 13). The native *S. vulgaris* causes some damage to conifers but this is not normally considered significant. *S. carolinensis* populations are managed at present mainly by baiting with warfarin, in combination with shooting and trapping. While these methods are adequately effective, the need to protect non-target animals and to avoid killing a pest which is also an aesthetically attractive animal is leading to consideration of other means of management (Kenward, this volume).

Glis glis, which is reported as a pest of forestry in Europe (Meehan, 1984), is restricted to a very small area around its site of introduction. It is reported from that area mainly as a surburban nuisance. This population should be monitored lest, like *S. carolinensis*, it spreads more widely and becomes a significant pest.

ACKNOWLEDGEMENTS

My sincere thanks to John Greaves, Adrian Meyer and Andy Swash for many comments and to numerous others for information on current damage levels.

REFERENCES

Bentley, E.W. (1964) A further loss of ground by *Rattus rattus* L. in the United Kingdom during 1956–61. *J. Anim. Ecol.*, **33**, 371–3.

Corbet, G.B. and Southern, G.B. (1977) *The Handbook of British Mammals*. 2nd edn. Blackwell, Oxford.

Drummond, D.C. (1985) Developing and monitoring urban rodent control programmes. *Acta Zool. Fenn.*, **173**, 145–8.

Drummond, D.C., Taylor, E.J. and Bond, M. (1977) Urban rat control: further experimental studies at Folkestone. *Environ. Health*, **85**, 265–7.

Dubock, A.C. and Kaukeinen, D.E. (1978) Brodifacoum (Talon rodenticide), a novel concept. *Proc. Eighth Vertebrate Pest Conf.*, Sacramento, California, March 1978 (ed. W.E. Howard) pp. 127–37.

Elton, C. (1942) *Voles, Mice and Lemmings: Problems in Population Dynamics*. Clarendon Press, Oxford.

Gratz, N.G. (1984) The global public health importance of rodents in *The Organisation and Practice of Vertebrate Pest Control* (ed. A.C. Dubock), ICI Plant Protection Division, pp. 413–35.

Greaves, J.H. (1985) The present status of resistance to anticoagulants in *Proc. Third Int. Theriological Congr.*, Helsinki, Finland (ed. A. Myllymäki) *Acta Zool. Fenn.*, **173**, 159–62.

Greaves, J.H. (1988) The control of commensal rodents in Britain. *EPPO Bulletin*, **18**(2), 203–9.

Greaves, J.H. and Cullen-Ayres, P.B. (1987) Genetics of difenacoum resistance in the rat in *Current Advances in Vitamin K Research. Proceedings of the 17th Steenbock Symposium*, Madison, Wisconsin (eds J.W. Suttie *et al.*), Elsevier, Amsterdam.

Greaves, J.H., Redfern, R., Ayres, P.B. and Gill, J.E. (1977) Warfarin resistance: a balanced polymorphism in the Norway rat. *Genet. Res. Camb.*, **30**, 257–63.

Greaves, J.H. and Rennison, B.D. (1973) Population aspects of warfarin resistance in the brown rat, *Rattus norvegicus. Mammal Rev.*, **3**, 27–9.

Greaves, J.H., Shepherd, D.S. and Quy, R. (1982) Field trials of second generation anticoagulants against difenacoum-resistant Norway rat populations. *J. Hyg. Camb.*, **89**, 295–301.

Green, R. (1979) The ecology of wood mice (*Apodemus sylvaticus*) on arable farmland. *J. Zool. Lond.*, **188**, 357–77.

Hardy, A.R. and Taylor, K.D. (1979) Radio tracking of *Rattus norvegicus* on farms. in *Handbook of Biotelemetry and Radio Tracking* (eds C.J. Amlaner and D.W. MacDonald), Pergamon Press, Oxford.

Huson, L.W. and Rennison, B.D. (1981) Seasonal variability in Norway rat (*Rattus norvegicus*) infestation of agricultural premises. *J. Zool. Lond.*, **194**, 257–89.

Lund, M. and Lodal, J. (1986) Resistance in the brown rat. *Danish Pest Inf. Lab. Ann. Rep.*, 1985, 84.

MacNicoll, A.D. and Gill, J.E. (1987) The occurrence and significance of rodenticide resistance in the UK in *Stored Products Pest Control*, BCPC Monograph 37, British Crop Protection Council, pp. 89–95.

MAFF (1975) *Pest Infestation Control Laboratory Report* 1971–73. Ministry of Agriculture, Fisheries and Food, Agricultural Development and Advisory Service. HMSO, London.

MAFF (1982) *Agricultural Science Service, Mammal and Bird Pests* 1981. Ministry of Agriculture, Fisheries and Foods, Agricultural Development and Advisory Service. HMSO, London.

Meehan, A.P. (1984) *Rats and Mice. Their Biology and Control.* Rentokil, East Grinstead.

Meyer, A.N. and Drummond, D.C. (1980) Improving rodent control strategies in Lambeth. *Environ.* Health, **88**, 77–81.

Myllymäki, A. (1979) Importance of small mammals as pests in agriculture and stored products in *Ecology of Small Mammals* (ed. D.M. Stoddart), Chapman and Hall, London, pp. 239–79.

Myllymäki, A. (1986a) Control of rodent problems by the use of rodenticides: rationale and constraints in *Control of Mammal Pests (eds C.G.J. Richards and T.Y. Ku), Trop. Pest Mgmt.*, **32**, Suppl. no. 1, pp. 83–111.

Myllymäki, A. (1986b) Efficacy of anticoagulants and flupropadine on wild strains of the house mouse, *Mus musculus* L., in Finland in *Proceedings of the Second Symposium on Recent Advances in Rodent Control*, Kuwait (eds A.H. Helmy Mohammad, T.M. Zaghloul, A.M. Salit and A.M. Zakaria), pp. 241–50.

Pelz, H-J. (1986) Damage by wood mice (*Apodemus sylvaticus*) to sugarbeet seeds in relation to food availability in *Control of Mammal Pests* (eds C.G.J. Richards and T.Y. Ku), *Trop. Pest Mgmt*, **32**, Suppl. no. 1, pp. 5–14.

Rennison, B.D. and Drummond, D.C. (1985) Monitoring and improving rodent control progress in non-agricultural premises in England and Wales. *Environ. Health*, **92**, 287–97.

Richards, C.G.J. (1985) The population dynamics of *Microtus agrestis* in Wytham, 1949 to 1978. *Acta Zool. Fenn.*, **173**, 35–8.

Richards, C.G.J. and Buckle, A.P. (1986) Towards integrated rodent pest management at the village level in *Control of Mammal Pests* (eds. C.G.J. Richards and T.Y. Ku), *Trop. Pest Mgmt.*, **32**, Suppl. no. 1, pp. 293–312.

Three

Ecological aspects of damage to sugar beet seeds by *Apodemus sylvaticus*

H-J. PELZ

SUMMARY

Damage by wood mice *Apodemus sylvaticus L.* to sugar beet seeds was first observed in the late 1960s and is now a common feature in western European sugar beet cultivation areas. Ecological factors of possible importance for the incidence of damage are discussed. Apparently damage occurs independently of the population dynamics of wood mice. Losses show some correlation with climatic factors. Availability of certain food components (invertebrates, cereals) is judged to be of prime importance with weather conditions acting indirectly by modifying the accessibility of these food items. Food shortage in spring might be enhanced by insecticide use. Diversionary feeding of wood mice with cereals between sowing and emergence of sugar beet seems to be a promising way to prevent losses.

INTRODUCTION

The wood mouse is one of the most common small mammals in Europe. This species can cope with a wide range of different habitats and dominates most biotopes (Niethammer, 1978). Despite their abundance wood mice have relatively rarely been shown to damage crops. In the past it was assumed that they could interfere with natural regeneration in forests by consumption of tree seeds. In contrast, Jensen (1985) demonstrated that wood mice and other rodents play a beneficial role in the rejuvenation of forests. By burying seeds they increase the chance of successful seedling formation.

The only field crop seriously threatened by *A. sylvaticus* is sugar beet. Feeding on other agricultural crops such as cereals is not of economic importance, as population density is moderate and outbreaks in numbers of the kind found in voles are not known for *Apodemus*. Minor losses due to wood mouse feeding can be ignored taking into account the beneficial effects resulting from the consumption of weed seeds and pest invertebrates by this species.

Damage to sugar beet seeds was first noticed in Lower Saxony in 1968 (Gersdorf, 1971) and was later observed in other parts of Europe, including the United Kingdom (Edgar and Isaacson, 1974), the Netherlands (Jobsen, 1983), Belgium, France, Italy and Denmark (personal communications).

Wood mice have a very fine sense of smell and are able to locate coated seeds buried in the soil. They dig out the coated seeds, gnaw off the coating, crack the pericarp and consume the embryo. The nocturnal wood mouse needs to take in 2–4 g of high energy food during one night. With one sugar beet embryo weighing only 3–4 mg a single mouse can cause serious damage if it feeds mainly on sugar beet seeds. Feeding tests in the laboratory with sugar beet seeds as the only source of food showed that one individual may consume on average 800 seeds per night. Maximum consumption by one individual was 1400 seeds in one night (Pelz, 1979).

Usually damage occurs all over the field with edges, and rows with seeds sown at lower depth or remaining at the surface being particularly at risk. Often wood mice follow a row and dig out all seeds for 10–20 m, then change rows. Damage shows up typically as small craters with remnants of seed coating and pericarp lying about.

The problem has been investigated in England (Green, 1979 and personal communication; MAFF, 1981) and West Germany (Pelz, 1979, 1986). Wood mouse population dynamics and incidence of losses seem to follow similar patterns in England and West Germany. There is some coincidence in the years concerned, but the extent of damage varies greatly between the two countries and between different areas within the countries (MAFF, 1981; Pelz, 1986).

In this paper I review the relevance of several factors of possible influence on the incidence of damage. Based on new data collected in recent years I try to assess relative importance and contribution of these factors to a situation where losses occur.

METHODS

Discussion of present knowledge of factors relevant to wood mouse damage is supplemented by new data. These data were collected as given below.

Field studies were conducted on farmland in the Rhineland (Figure 3.1) in 1976 and 1977 and from 1980 to 1986. During this time 465 wood mice were caught in snap traps. Sex, weight and reproductive state were recorded for each individual. Relative importance of food components was assessed by stomach contents analysis as described by Hansson (1970). Palatability of different food items was checked in cages in the

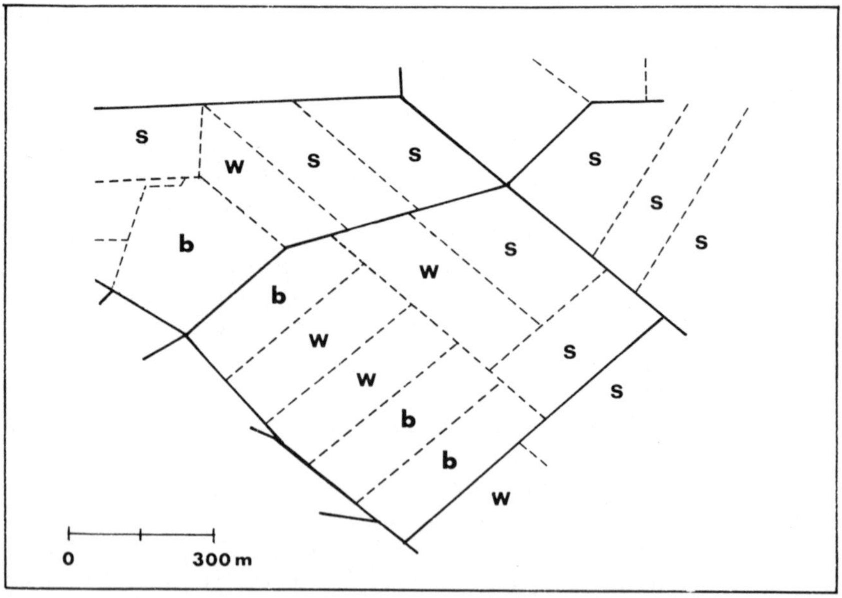

Figure 3.1 Map of the study area in the Rhineland in spring 1984. Solid lines – field paths; broken lines – field boundaries; b – winter barley; s – sugar beet; w – winter wheat.

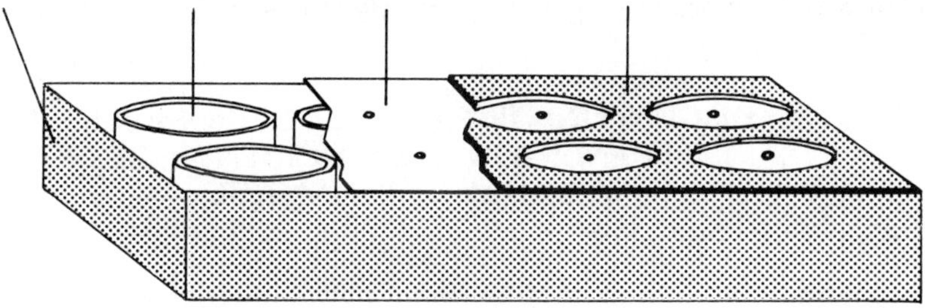

Figure 3.2 Food preference test box. For explanation see text.

laboratory. These tests were carried out with a special test-box (Figure 3.2). By sniffing at the holes and gnawing the cardboard wood mice could access and feed on a specific food item. No other food was available during the test. Each collection of food items was tested with six mice during one week. Evaluation was based on means of choice (gnawing off the cardboard above a special food item) and amount consumed. According to the results of food-choice tests items were put into a rank order of preference.

Influence of precipitation on the ability of wood mice to find and dig out pelleted sugar beet seeds was investigated under well-defined conditions in indoor pens. Before the start of the activity-phase of the wood mouse living in that pen, the soil was prepared by sowing 72 seed pills 1 cm deep on an area of 2 m^2. The respective amount of water was evenly sprayed on the dry ground after sugar beet seeds had been sown. The number of seeds taken was determined after 5.30 h.

Incidence and severity of damage to sugar beet seeds in the field was judged by survey results from 1982 onwards. Sugar beet growers in several areas of West Germany were asked about occurrence of damage, time of sowing and germination, use of pesticides or other measures of agricultural practice, and precipitation during the sowing period. A mean number of 70 fields was included in the survey each year with an area of 335 ha on average. A four-point scale of damage severity was used:

1. Very low; damage occurs sporadically, without economic importance.
2. Low; little but noticeable damage, resowing is not necessary.
3. Medium; considerable damage in parts of fields, partial resowing may be necessary.
4. High; total area is heavily damaged, new sowing is necessary.

Relative abundance of wood mice was calculated on the basis of the results of snap trap catches performed by the farmers to control wood mice. These trap lines were placed on the edge of sugar beet fields about 40 m apart, before sowing or immediately after sugar beet had been sown.

DISCUSSION OF FACTORS OF POSSIBLE INFLUENCE ON THE INCIDENCE OF WOOD MOUSE DAMAGE

General background

Sugar beet is among the most profitable of field crops and thus is generally grown on the most fertile soils. In sugar beet areas hedgerows and trees have been cleared out almost completely. This is the case in most parts of western and central Europe, the only exception being windy areas where shelterbelts are needed.

The data on wood mouse ecology discussed in this paper were gathered in a typical sugar beet growing area in the Rhineland between the cities of Cologne and Aachen (Figure 3.1). Average field size in that area is 4–5 ha, average farm size 30–40 ha. Bushes and trees are completely absent (except some around farmhouses), the nearest wood is present at a distance of eight kilometres. Farmers practise a three-year crop rotation with sugar beet – winter wheat – winter barley. Sugar beet is sown between mid-March and mid-May (the majority in April), winter barley or rye in September/October, winter wheat between mid-September and mid-December, dependent on weather conditions.

Wood mice are found in this type of farmland away from cover throughout the year (Pelz, 1979). Since field habitats change with season and agricultural measures, *A. sylvaticus* is confronted with new conditions at least twice during the year. With decreasing habitat quality woodmice populations migrate to more favourable sites. Peak population densities can be observed in autumn before harvest in sugar beet fields. After harvest of sugar beet and ploughing most wood mice migrate into adjacent winter barley fields. In spring the species is most abundant in wheat (Pelz, 1979).

Sugar beet

Ordinary sugar beet 'seeds' are multigerm, with a pericarp of irregular shape containing several embryos. Sugar beet seeds are sown at a depth of 1–3 cm in rows 45–50 cm apart. To achieve an optimum density of 75 000 plants/ha, seeds sown with a spacing of 4–8 cm have to be thinned out by hand or machine to result in a 25 cm spacing between plants. Modern sowing techniques with planting-to-stand were achieved by

Table 3.1 Development of application and drilling width of genetically monogerm seed pills in the catchment area of the sugar factory of Ameln, Rhineland

Year	Monogerm seed (% of total sugar beet area)	Mean seed spacing (cm)
1969	—	8
1970	2.63	13.5
1971	4.58	18.0
1972	14.09	16.4
1973	26.15	14.7
1974	43.10	14.4
1975	60.90	14.0
1976	76.93	13.9
1977	87.98	13.5
1978	91.67	13.9
1979	95.37	14.8
1980	97.70	16.2
1981	98.98	18.1

Data: Pfeifer and Langen, Werk Ameln

breeding monogerm 'seeds' and giving the seeds a smooth coating to form regular seed-pills. The coating may contain fungicides and insecticides to prevent losses due to soil-borne diseases. These seed-pills can be sown with a precision drill and there is no need for later thinning of seedlings. Table 3.1 presents data on the increase of monogerm seed use in the catchment area of a sugar beet factory in the Rhineland. This area suffered severe wood mouse damage for the first time in 1976, when nearly 80% was sown with monogerm seed. The importance of monogerm seed use for the incidence (or recognition?) of damage becomes apparent by comparison of the situation in the Rhineland with regions, where (due to certain socioeconomic or climatic conditions) planting-to-stand with monogerm seed is less common. For instance, in Bavaria the share of monogerm seed in 1982 was still below 50% (Schneider and Ortmaier, 1984), and wood mouse damage is nearly unknown in that area. With increasing use of monogerm seed damage in all western European countries increased, and first became of economic significance in the middle of the 1970s with losses up to 10% of seeds. A loss of 10% requires resowing.

Wood mouse populations

Wood mice have few competitors in farmland habitats; only occasionally some specimens of *Microtus arvalis* or *Mus domesticus* may be present.

Table 3.2 Average precipitation, wood mouse abundance and damage in March/April from 1982 to 1987

Year	Damage assessment	Relative abundance* of *A. sylvaticus* (%)	Rainfall/24 h† (mm)
1982	Medium/high	75	1.2
1983	Very low	203	3.0
1984	High	121	1.0
1985	Low	59	1.7
1986	Low	87	1.5
1987‡	Low	55	1.7

* Relative abundance was calculated from snap trappings covering field areas of 170 ha on average/year. The average wood mouse density 1982–1987 was 1.14 individuals/ha.

$$\text{Relative abundance} = \frac{\text{number of woodmice/ha} \times 100\%}{\text{average number of wood mice/ha}}$$

† Data for the period from sowing to emergence.
‡ Preliminary.

Natural enemies are predatory mammals and nocturnal raptors. For much of the year fields offer little cover and owls could be efficient predators. Nevertheless, the most efficient one, the barn owl, *Tyto alba*, is now very rare.

Population density of wood mice on farmland and in woodland areas follows similar patterns. There are pronounced annual fluctuations with highest densities in autumn/early winter and lowest densities in late spring/early summer (Green, 1979; Pelz, 1979). On farmland peak densities are observed in sugar beet fields (around 40–80 individuals per ha). Ploughing causes a sharp population decline. Until March/April, when sowing of sugar beets starts, numbers are further reduced to 0.5–8 individuals per ha (Pelz, 1979).

The population density of wood mice seems to have little effect on seed damage (Table 3.2). Population density in spring was high in 1983, which was the year with the lowest incidence of damage. In contrast, in 1982 and 1984 losses were high in spite of relatively low population density.

Body weight and reproductive state were used as indicators for the condition of the population. Data from years without losses were compared with data from years where damage did occur. There were no significant differences between years with losses and without losses (Pelz, unpublished).

Diet of wood mice in beet-growing areas

Sugar beet seeds are taken while wood mice are on their nocturnal search for food. Figure 3.3 summarizes results of stomach contents analyses over

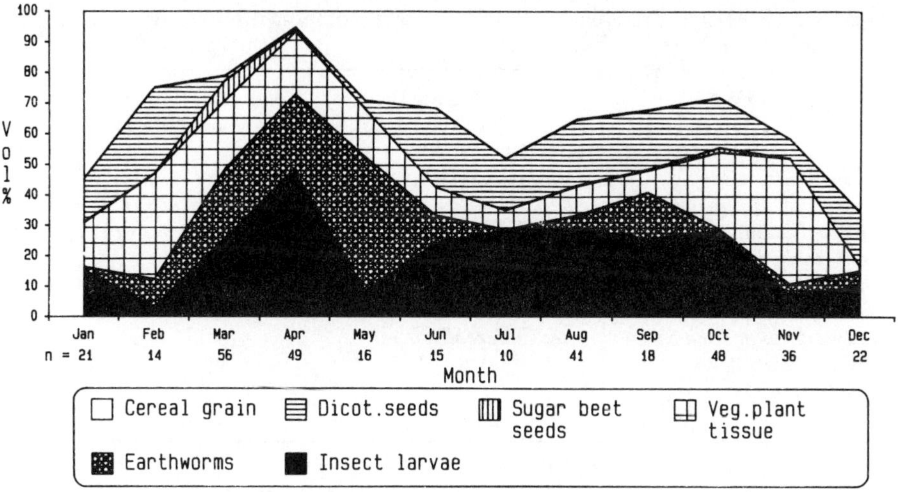

Figure 3.3 Diet of wood mice on arable farmland (analysis of stomach contents, pooled data of several years).

the year. Cereals and invertebrates are the basic food components. Weed seeds and vegetative parts of plants supplement the diet, according to season. The high amount of invertebrates consumed in spring is remarkable. Statistical analysis of these data demonstrated no difference in diet composition between the sexes.

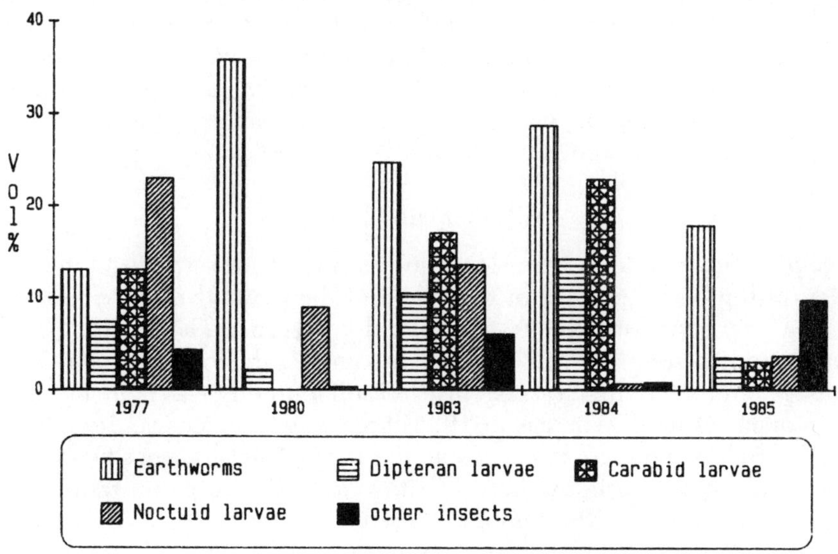

Figure 3.4 Invertebrates in the diet of wood mice on arable farmland.

Table 3.3 Relative preference of some dietary components available on arable farmland (results of food choice tests)

Group A high preference	Group B low preference	Group C not consumed
*Helianthus annuus**	*Viola arvensis**	*Thlaspi arvense**
Alopecurus myosuroides†	*Rumex obtusifolius**	*Matricaria chamomilla**
Hordeum vulgaris†	*Lumbricus terrestris*	*Plantago major**
Scotia segetum§	*Sinapis alba**	
Beta vulgaris‡	*Chenopodium album**	
Triticum sativum†	*Capsella bursa-pastoris**	
*Polygonum persicaria**	*Secale cereale*†	
*Beta vulgaris***		
*Zea mays**		
Avena sativa†		

* Seeds, † caryopses, ‡ seed pills, § larvae, ** root tissue.
For groups A and B results are given in a provisional order of preference.

With respect to invertebrate food (Figure 3.4) the amount of earthworms is the most constant element. Consumption of insect larvae varies strongly in amount and insect families concerned, probably due to changes of abundance of the insect species in different years.

Table 3.3 summarizes the results of palatability tests with food components in the laboratory. Items are classified into three groups according to relative preference. Cereals, insect larvae and sugar beet seeds were found to be highly preferred. Earthworms were less often consumed and seem to be of medium preference for wood mice. In other tests, with sugar beet seeds available *ad libitum*, a strong negative correlation between the availability of insect larvae and the amount of sugar beet seeds consumed was found (Pelz, 1986).

Weather conditions

Precipitation. The survey on losses and rainfall demonstrates an inverse relationship between level of damage and the amount of precipitation during the period between sowing and emergence (Table 3.2). This correlation is very obvious in comparison of data for 1983 (minimal damage) and 1984 (high losses). In 1983 soil humidity was high, because of frequent rainfall, whereas in 1984 there was an extensive dry period that led to low moisture levels in soil. For the period between sowing and emergence years without damage show a higher level of rainfall in contrast to years with damage (Figure 3.5). However, in tests in enclosures with controlled precipitation, only a weak relationship between the amount of rainfall and the ability of wood mice to discover

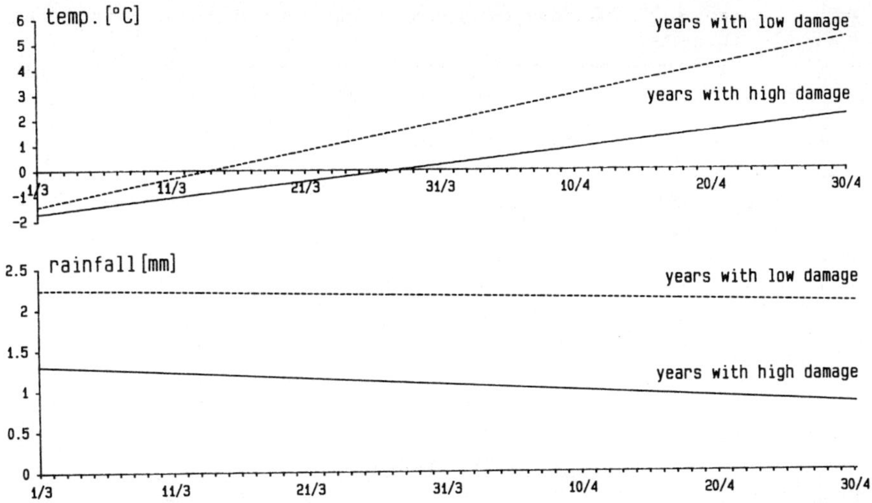

Figure 3.5 Comparison of years with low and high wood mouse damage: trends of minimum temperature at ground level and rainfall in March/April.

sugar beet seeds could be found (Figure 3.6). Even with a precipitation of 32 mm/m^2, which is very unlikely to occur with one cloudburst in the field, both test animals found more than 60% of the seeds in the soil. Further experiments should confirm whether this applies also for heavier and more loamy soils. The results of the survey suggest, however, that damage occurs regardless of type of soil (Table 3.4).

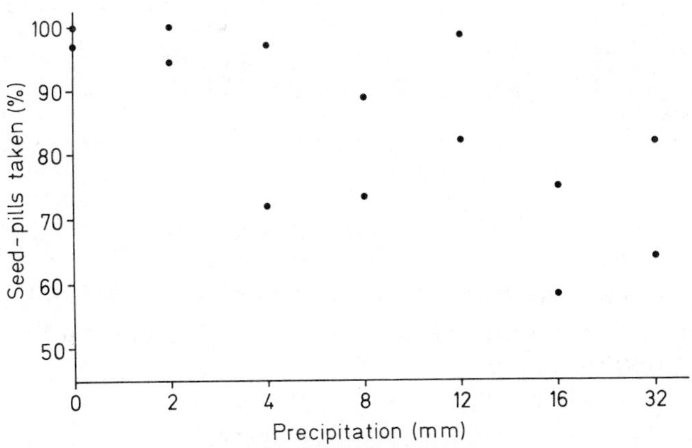

Figure 3.6 Correlation between precipitation and wood mice digging out sugar beet seeds. Tests with light soil in indoor pens at 25°C with two replicates at each level of precipitation ($y = 93.089 - 0.81x$, $r^2 = 0.36$, $p \leq 0.05$).

Table 3.4 Correlation between wood mouse damage and type of soil

Soil type	Total number of fields	Number of fields damaged
Light	36	6 (16.6%)
Medium	149	25 (16.8%)
Heavy	42	7 (16.7%)
Total	227	38 (16.7%)
Sand	33	8 (24.2%)
Loam	133	18 (13.5%)
Total	166	26 (15.7%)

Temperature

Similar tendencies showed up with temperature. Figure 3.5 gives trends of minimum temperature on soil surface in years with and without damage. The critical time for losses due to wood mice is between sowing and emergence. After emergence sugar beet seeds are rarely taken by wood mice. However, according to Table 3.5 it is not a delay of germination due to cold weather that increases severity of losses.

Table 3.5 Comparison of mean time from sowing to emergence for sugar beet fields with and without losses due to wood mouse feeding (data 1982–1987)

	n	Mean time sowing–emergence (days)	*s.d.*
Fields without damage	222	14.03	±4.31
Fields with damage	52	14.13	±4.92

PREVENTIVE MEASURES

Ways by which farmers could possibly influence the probability of damage to sugar beet seed include choice of sowing time, use of pesticides and application of repellents.

Sowing time

Generally sugar beet is sown as early as possible in spring to obtain maximum yield, but since the seedling is tender, there is a higher risk of losses due to bad weather conditions (e.g. frost, heavy rainfall encrusting the soil surface) when sowing starts very early. Time from sowing to emergence is about 10 days under good conditions (e.g. sufficient soil temperature and moisture), but can take more than double that time at low temperatures or in dry weather. Results of our survey suggest that there is no general time pattern for the incidence of damage by wood mice (Pelz, 1986). In 1984, when severe losses occurred, damage was observed at a similar level independent of sowing time. During the six years of the survey only fields that were sown before 20 March or after 30 April had never been damaged, but these were no more than 13.6% of all fields considered.

There are two reasons why wood mouse damage is less likely with very early sowing dates:

1. With freezing temperatures at night, the soil surface will be frozen and thus sugar beet seeds cannot so easily be dug out.
2. Range and activity of wood mice are reduced during winter (Gurnell, 1975; Green, 1979).

If sowing date is very late, early-seeding weeds (e.g. *Stellaria media*) may already be available.

Use of pesticides

Insecticides containing lindane are frequently applied before sowing. They can be sprayed or applied as a granular compound. Another pesticide frequently applied as granules to the seed furrow is aldicarb, a nematicide and insecticide.

Table 3.6 Incidence of wood mouse damage in fields with and without lindane treatment

Treatment	Total number of fields	Number of fields damaged
Lindane	101	19 (18.8%)
No lindane	39	2 (5.1%)

The survey on losses due to wood mice includes information on insecticide use from 1985 onwards (Table 3.6). There appears to be a correlation between the application of insecticide and the incidence of damage, but since no heavy damage occurred during the years surveyed, these data can only give some preliminary indications.

Repellents

Coating seeds offers the opportunity to add substances to the seed-pills that might act as a repellent. During the 1970s tests with many substances were conducted in England (Dunning and Byford, 1978) and West Germany (Grimm and Range, 1976). Pelz (1979) tested some pesticides and about 150 odour compounds from the cosmetic industry using the method described above for food preference tests. Although most of the substances tested proved ineffective, there were a few promising results, but no compound has so far led to practical application because either of problems with phytotoxicity or lack of cost-effectiveness.

CONCLUSIONS

Several factors seem to contribute to conditions causing losses by wood mice in sugar beet fields. Possible interactions of these factors are shown diagrammatically in Figure 3.7. Wood mice are active in sugar beet fields during their nocturnal search for food. Consumption of a certain food

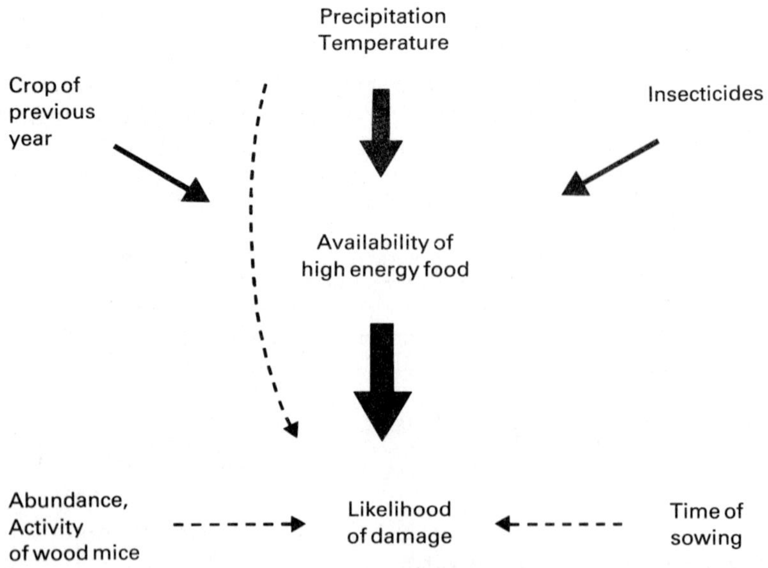

Figure 3.7 Simplified scheme of relevant factors affecting the incidence of wood mouse damage.

item is primarily dependent on the amount of that food present. The main source of cereal and other seeds is from the crop of the previous year. The abundance of invertebrates, the most important food component during sowing time, depends on the population dynamics of the insect species concerned, and their accessibility at the soil surface depends on soil temperature and soil moisture. This might explain the increase of wood mouse damage in periods of dry and cool weather. Treatment with insecticides possibly increases sugar beet losses due to wood mice by reducing this invertebrate food source.

The abundance and activity of wood mice seem to be of minor importance to levels of damage. As it is difficult to predict availability of insects, or weather conditions, there seems to be no means at the moment to predict levels of damage. Thus, it will be necessary for farmers to take preventive measures each spring.

The importance of food availability during sowing of sugar beet is illustrated by the fact that wood mice readily consume cereals in small boxes, which are provided along the field as a diversionary tactic (Pelz, 1985). The attractiveness of these feeding stations positioned on the edge of newly sown sugar beet fields 30 m apart was so high that some individuals moved into the boxes and could be found there during daytime. This diversionary tactic of providing wood mice with alternative food during the critical time after sowing has been tested by several farmers during recent years. Schulze-Lammers (1986) threw a handfull of barley seeds every 30 m from a moving car on the edge of the newly sown sugar beet fields. He found it unnecessary to prevent the barley from germination since all grain disappeared due to feeding of mice and presumably birds.

No losses have been found so far on fields provided with alternative food. As losses occur irregularly and locally, lack of damage is not a definite proof of the efficacy of this technique. Further trials are required and additional methods of providing food that fit in with local agricultural practice should be tested. However, the idea of diversionary feeding seems to be very promising, as it is cheap, easily done and an ecologically beneficial way to prevent losses by wood mice.

ACKNOWLEDGEMENTS

I wish to thank R.J. Putman and two referees for helpful comments on an earlier draft of this paper and for revising the English text. Some of the work was supported by Deutsche Forschungsgemeinschaft (Pe 281/1-1). I am particularly grateful to I. Burfeindt and H. Naujeck for their assistance with research reported here.

REFERENCES

Dunning, R.A. and Byford, W.J. (1978) Sugar beet seed treatments in CIPAC Monograph 2, *Seed Treatment* (ed. K.A. Jeffs), Cambridge, pp. 79–90.

Edgar, W.H. and Isaacson, A.J. (1974) Observations on skylark damage to sugar-beet and lettuce seedlings in East Anglia. *Ann. Appl. Biol.*, **76**, 335–7.

Gersdorf, E. (1971) Waldmäuse und Vögel zerstören pillierte Rübensaat. *Nachrichtenblatt des Deutschen Pflanzenschutzdienstes*, **23**, 151–2.

Green, R. (1979) The ecology of wood mice (*Apodemus sylvaticus*) on arable farmland. *J. Zool.*, **188**, 357–77.

Grimm, H. and Range, H. (1976) Waldmäuse (*Apodemus sylvaticus* L.) als Schädlinge im vereinzelungslosen Rübenanbau. Rundbriefe Ackerbau, SUET: Saat- und Erntetechnik, D-3440 Eschwege.

Gurnell, J. (1975) Notes on the activity of wild wood mice, *Apodemus sylvaticus*, in artificial enclosures. *J. Zool.*, **175**, 219–29.

Hansson, L. (1970) Methods of morphological diet micro-analysis in rodents. *Oikos*, **21**, 255–66.

Jensen, T.S. (1985) Voles and mice as seed dispersal agents. *Fourth International Theriological Congress, Edmonton* 1985, Abstract no. 324.

Jobsen, J.A. (1983) Schade aan pas gezaaide bieten door de Boismuis. *Maanblad Suiker Unie*, **17**, 15.

MAFF (1981) *Field Mice and Sugar Beet.* Ministry of Agriculture, Fisheries and Food. Leaflet 626, 6 pp.

Niethammer, J. (1978) *Apodemus sylvaticus* (Linnaeus 1758)–Waldmaus in *Handbuch der Säugetiere Europas*, Band 1, *Rodentia* I (eds J. Niethammer and F. Krapp), Akademische Verlagsgesellschaft, Wiesbaden, pp. 337–58.

Pelz, H-J. (1979) Die Waldmaus, *Apodemus sylvaticus* L. auf Ackerflächen: Populationsdynamik, Saatschäden und Abwehrmöglichkeiten. *Z. Angew. Zool.*, **66**, 261–80.

Pelz, H-J. (1985) Zur Prognose und Abwehr von Waldmausschäden an Zuckerrübensaat. *Gesunde Pflanzen*, **37**, 224–9.

Pelz, H-J. (1986) Damage by wood mice (*Apodemus sylvaticus*) to sugar beet seeds in relation to food availability in *Control of Mammal Pests* (eds C.G.J. Richards and T.Y. Ku), *Trop. Pest Mgmt*, **32**, Suppl. 1, 5–14.

Schneider, K. and Ortmaier, E. (1984) Betriebswirtschaftliche Bedeutung des Zuckerrübenanbaues am Beispiel der Verhältnisse im süddeutschen Raum in *Geschichte der Zuckerrübe.* 200 Jahre Anbau und Züchtung, Berlin, pp. 152–174.

Schulze-Lammers, H. (1986) Erfahrungen aus der Praxis. Pflanzenschutz-Praxis, 21–22.

Four

Fossorial voles: problems and research

A. MEYLAN

The majority of voles have a subterranean shelter made up of galleries connected to one, or several, nests; yet most of these small mammals move on the soil surface and feed on epigeal vegetation. There are, however, some species, the home ranges of which are almost exclusively limited to underground gallery systems and which feed on plant roots.

In Europe, the fossorial form of the water vole, *Arvicola terrestris scherman*, and the voles belonging to the genus *Pitymys* show such characteristics. Damage caused by subterranean voles is acute, since plants whose root systems have been destroyed do not regenerate, in contrast to surface-grazed plants. The importance of these injuries depends on the crop type and on the pest density.

As a result of the behaviour of the fossorial voles, the control of populations is difficult. On the one hand, voles may go unobserved because they are occupying galleries dug by moles, genus *Talpa*. On the other hand, the efficiency of control techniques, which need to act in their burrows, is not always satisfactory.

During the past twenty years, the fossorial form of the water vole, which occurs in Central Europe, has been the subject of much research. Starting from sampling by trapping out animals, studies have been carried out by capture-mark-recapture, radioactive tagging and radiotracking. If, up to now, characteristics of space occupation and movements, of social relationships and activity rhythms of this large rodent are relatively well known, factors which influence the level of population densities during cyclic outbreaks still remain to be understood. Investigations on the small voles of the genus *Pitymys*, with more than ten different species occurring in Europe, are just beginning, but preliminary results show a great diversity in their biology.

A better understanding of the ecology of fossorial voles is necessary to develop new control strategies, which involve not only suitable and efficient poisoning methods but also habitat manipulations and the protection of predators.

Five

Prevalence of *Pneumocystis carinii* and *Leptospira icterohaemorrhagiae* in Danish rodents

J. LODAL AND M. LUND

In many areas rats and mice are serious pest animals because of the economic losses they cause to growing crops and stored products. Another pest aspect is their potential role of spreading disease to humans and domestic animals.

Pneumocystis carinii pneumonia and Weil's disease caused by *Leptospira icterohaemorrhagiae* are two serious infections, both of which may be fatal to infected humans. In collaboration with medical institutes we have studied the prevalence of *P. carinii* and *L. icterohaemorrhagiae* in Danish rodents.

P. carinii is an enigmatic organism now believed to be a protozoan. Its mode of transmission is unknown. There are indications of *P. carinii* being transmitted from rodents to humans by air. *P. carinii* was found in 17 of 90 (18.9%) of wild-living trapped *Rattus norvegicus*. The positive rats originated from about 25% of the trapping localities examined. Furthermore, among wild-living trapped animals one of 16 *Mus musculus* and one of eight *Apodemus sylvaticus* were found positive. Small numbers of *Apodemus flavicollis*, *Clethrionomys glareolus*, *Microtus agrestis* and *Arvicola terrestris* were also examined but all of them were negative. One of ten *Rattus rattus* of a laboratory breeding stock was found positive. Among rodents in Denmark *Rattus norvegicus* seems to be the most important reservoir host for *Pneumocystis carinii*.

Weil's disease is caused by the spirochaete *L. icterohaemorrhagiae*. There are few human deaths each year in Denmark but there has been a preponderance of fish pond workers among the infected patients. The brown rat, *Rattus norvegicus*, is known to be an important animal carrier of the organism which is excreted with the urine. A study was, therefore, undertaken in 1985 and 1986 to elucidate whether leptospires were more frequent among rats from fish pond areas than in rats from agricultural premises. A total of 118 rats were trapped at 38 fish ponds and 106 rats at 34 agricultural premises. They were all examined for leptospires in their kidneys and *L. icterohaemorrhagiae* was found in four rats from fish ponds and in three from agricultural premises. Positive rats were found only in 1985. The proportion of fish ponds having infected rats was 28.6% in 1985 compared to 0 in 1986, and the proportion of agricultural premises was 30% in 1985 compared to 0 in 1986. Consequently, the higher frequency of leptospirosis among fish pond workers than among other occupational groups must be due to factors other than the proportion of infected rats.

Six

Progress in rodent control and strategies for the future

A.B. LAZARUS

SUMMARY

The control of rodents, which are amongst the most significant pest species in terms of economic damage and disease transmission, relies primarily on the use of toxic chemicals. Trapping, although effective against small, discrete infestations, is heavily labour intensive, and the use of ultrasonics or other repellent devices is not of proven effectiveness. The continued development and spread of anticoagulant resistance, particularly to the older hydroxycoumarins (such as warfarin) and indandiones, has meant greater reliance on the use of rodenticides with high single-dose toxicity and these include the newer, more potent, anticoagulants. The greater risks to non-target species and potential for damaging the environment posed by these poisons must be counteracted by updating the methods of rodenticide formulation and application in order to target more specifically the infestation to be controlled. A greater commitment of resources to research in rodent control could result in the early development of species-specific rodenticides. The incorporation of additives to bait and contact formulations might render rodenticides less harmful or otherwise unattractive to non-target species. Novel methods of rodent control, including the use of pheromones and reproduction inhibitors, could also be incorporated into control programmes. Co-ordinated area programmes or 'integrated control' in which rodent control agents are skilfully utilized can eradicate rodent infestations and should prove to be most effective at preventing the further spread of rodenticide resistance.

INTRODUCTION

The Rodentia are by far the most numerous of all mammals, whether considered in terms of sheer numbers, numbers of species (of which there are over 6000) or variety in body form or function. The relatively unspecialized family Muridae contains many rodent pest species including the three cosmopolitan commensals – *Rattus norvegicus* (Berkenhout), *Rattus rattus* (Linnaeus) and *Mus domesticus* (Rutty). These pest species, of highly significant economic and public health importance are extremely adaptable and have high levels of fecundity and reproductive output.

Rodents cause major damage to a large variety of growing crops before harvest but equally important are the losses in value caused to harvested crops and packaging arising from rodent infestation (Berry and Jakobson, 1975; Drummond, 1971, 1974; Girish and Arora, 1976; Wilson, 1972). Rodents by their constant gnawing can damage the fabric of buildings, insulation, water pipes and electrical wiring and may be a major cause of many unexplained fires in farm buildings (Franco and Stazzi, 1970; Marsdin, 1975). The economic importance of rodent pest species is at present impossible to quantify; however, estimates of the annual costs of rodent control (largely unsubstantiated) include \$100 million in the USA during the early 1970s and £50 million in the UK during the early 1980s (Meehan, 1984).

Rodents act as reservoirs to many virulent diseases transmissible to man and to domestic and farm animals, thereby lowering their yield. Rodent-borne diseases include plague, the various typhuses and rickettsiae, schistosomiasis, leishmaniasis, dengue and lassa fevers and leptospiral infections among others (Bellier, 1973; Gratz, 1984).

METHODS OF RODENT CONTROL

Exclusion or deterrent devices

The proofing of buildings against entry by rodents and basic attention to hygiene can reduce and indeed have reduced much of the damage caused to stored products and the spread of disease by rodents. By contrast the many exclusion and deterrent devices employed against rodents including ultrasonic generators, ultraviolet lamps, magnetic devices, chemical repellents and electric fences (Howard, 1968; Kaufman and Michaelson, 1974; Meehan, 1984; Wilson, 1960) have produced, at best, variable results. Few have so far been of much practical value.

Trapping

Trapping must be amongst the oldest of the control methods employed against rodents as can be testified by the many traditional live and kill traps in different parts of the world. These include pitfall, deadfall, snare and the various types of cage traps. There are pottery traps in the museums of several Middle East countries that date back to the third millenium BC (Drummond, 1980).

Trapping can be used to reduce the numbers of rodents in an infestation, but it is very labour intensive and although multi-capture traps are used for mouse control in the USA (Jackson, 1987) trapping would be an unlikely method to contemplate as first choice for controlling large infestations. However, the use of snap or breakback traps can occasionally be the preferred method for controlling small populations of rats or mice in an urban environment or in food stores. Traps are mainly used these days by zoologists for collecting research or museum specimens or for ecological studies.

Sticky boards, although frequently used to entrap mice in the USA and other countries are not generally used in the UK because there are equally effective methods available that do not produce the same adverse reactions from the public (Baker, 1983; Frantz and Padula, 1983).

Other non-chemical control methods

Although predators including various species of mongoose and weasel and the domestic cat and dog have been used to reduce rodent infestations, they are of very limited effectiveness. This is partly because predators are unlikely to achieve large-scale reductions in rodent infestations and the introduction of predators, like the mongoose, can have disastrous consequences for non-target species (Wodzicki, 1973).

CHEMICAL METHODS OF RODENT CONTROL

The incorporation of chemicals in rodenticidal baits offers the most widely used and efficient methods for controlling rodent infestations. However, application of a rodenticide in water against rodents in dry areas or as a fumigant or in contact formulations against difficult-to-bait infestations are viable alternative control measures (Marsh, 1983; Meehan, 1984). The poisons used as rodenticides are rarely specific to the target species and therefore require great care in their application to prevent the poisoning of other species.

Single-dose non-anticoagulant (or acute) rodenticides

One of the oldest of the rodenticides documented is red squill, a scillirocide, which is essentially the dry powdered fleshy inner bulb scales of the red *Urginea maritima* plant. Its rodenticidal properties were known in the Mediterranean area in mediaeval times (Gratz, 1973). Red squill has limited acceptance by many animals other than rodents because it induces vomiting, whereas rodents cannot vomit and so cannot eliminate the poison. However, its efficacy as a rodenticide is extremely variable because of difficulties in obtaining a standard formulation.

Other single-dose rodenticides, including barium carbonate, fluoroacetamide, sodium fluoroacetate, yellow phosphorus, strychnine, thallium sulphate and zinc phosphide, have in the past, enjoyed widespread use (Freeman, 1954; Mallis, 1982). Of these only fluoracetamide and zinc phosphide are at present registered for use in the United Kingdom, together with the more recent single-dose rodenticides, alphachloralose and norbormide. In addition there are the cyanide and phosphide compounds that are used for toxic gas production during the fumigation of burrows. Typically, single-dose rodenticides (formerly known as acute poisons) act rapidly, producing symptoms of poisoning often within one hour and death within a few hours when a lethal dose has been consumed. The rapid onset of poison symptoms, however, together with neophobia often result in sublethal dosing and subsequent bait shyness so that 100% mortality is rarely achieved despite the use of elaborate baiting techniques. The non-selective toxicity of most single dose rodenticides together with the rapidity of poison action make them extremely hazardous to use and efficient application requires the services of skilled personnel. These factors partly account for the dramatic and sustained decline in their use following the introduction of the 'first generation' anticoagulants in the early 1950s (Crabtree, 1950).

Multiple-dose (or chronic) rodenticides

The introduction of warfarin and the other 'first generation' hydroxycoumarin and indandione anticoagulants revolutionized rodent control and made it possible to simplify rodent baiting techniques. The slow onset of poison symptoms (over several days) and the multiple intakes of bait necessary to achieve a fatal dose means that these poisons, to a large extent, act as their own prebait. No prebaiting, therefore, was necessary. The relative safety of these anticoagulants to non-targets, together with the increased efficacy and ready availability of an antidote in vitamin K_1, resulted in the almost universal reliance on them for tackling rodent problems.

It therefore came as a major shock when the resistance of *Rattus norvegicus* to warfarin and the other anticoagulants was first recorded in 1958 in Scotland (Boyle, 1960). Shortly afterwards warfarin resistance was also recorded in the Mid-Wales area. Subsequently, cross-resistance in *Rattus norvegicus* to most of the anticoagulants has been recorded, not only in this country, but also overseas (Greaves, Shepherd and Gill, 1982; Lund, 1982; Redfern and Gill, 1978). Several other rodent pest species were also found to be more tolerant of the anticoagulant rodenticides. Attempts to eradicate resistant rat populations with acute poisons were largely unsuccessful (Drummond and Bentley, 1967) although this method of control apparently worked in Holland (Ophof and Langeveld, 1969).

RESEARCH EFFORTS IN THE 1960s AND 1970s

The realization that large numbers of rodent infestations could no longer be controlled by the available anticoagulants provided the impetus for increased research effort into rodent control measures worldwide. Through the Tolworth Laboratory, MAFF developed close liaisons with the pesticide and pharmaceutical industries and with other research establishments to look, not only at previously used rodenticides but at the potential of new rodent poisons (Greaves, 1971). In addition, new rodenticide bait formulations, the potential of reproduction inhibitors, pheromones and ultrasonics as control methods were also investigated. Advances by the pesticide industry during this period included the development of new single-dose rodenticides such as alphachloralose (a narcotizing agent) and norbormide (which has a high degree of specificity for *Rattus norvegicus*) as well as calciferol (an effective multiple-dose non-anticoagulant rodenticide). These were developed in order to increase the range of effective rodenticides in an attempt to reduce the reliance on the anticoagulants.

Chemosterilants

Among the alternative methods investigated with interesting potential in rodent control was the use of chemosterilants. The concept for the regulation of reproduction in rodents and other pest species was first mooted by Knipling (1959). Davis (1961) conducted one of the earliest studies of the chemosterilant approach in rodent control and this approach has been subsequently investigated not only in rats (Kendle *et al.*, 1973; Marsh and Howard, 1969) but in several other species (Balser, 1964; Linhart and Enders, 1964; McDonald, 1980; Plant, 1983). Of the reproduction inhibitors investigated, several were synthetic oestrogenic steroids. One cyclopentyl oestrogenic steroid investigated at Tolworth was field tested on a rat-infested rubbish tip in West Sussex. Application of the chemosterilant at 0.05% in bait suppressed reproduction in rats occupying the tip for more than six months (Rowe and Lazarus, 1974). A similar application against the rat population occupying a small island in Poole Harbour inhibited reproduction for about one year and resulted in a significant decline in rat numbers (Lazarus and Rowe, 1982). Despite such promising potential, chemosterilants were never fully exploited as rodent control agents, although one non-steroidal compound, alpha-chlorohydrin, has been marketed both as a male-rat-specific chemosterilant and as a lethal rodenticide – a 'toxicant-sterilant' (Ericsson, 1982). However, alpha-chlorohydrin is not palatable to rats and is not a species-specific poison.

'Second-generation' anticoagulants

Several of the anticoagulants developed during the 1970s and 1980s have been found to be effective against warfarin-resistant rat populations and were heralded as the answer to 'resistance' problems. These 'second-generation' anticoagulants include difenacoum, bromadiolone, brodifacoum and flocoumafen (Smith and Greaves, 1986). The more potent of these anticoagulants are effectively single-dose poisons, although the toxic symptoms in rodents are delayed for several days and control may not be obtained any sooner than with warfarin. The 'second-generation' anticolagulants are relatively expensive chemicals, however, and generally present greater potential hazards to non-target species because of their increased toxicity and persistence. The more toxic of these, brodifacoum and flocoumafen (Dubock, 1980; Garforth and Johnson, 1987) are yet to be cleared for general or outdoor use against rodent infestations in the UK.

The relatively high cost, greater toxicity and single-dose effectiveness of the 'second generation' anticoagulants have resulted in the advocation of the restricted or 'pulsed' baiting concept for their application (Dubock, 1982). Baiting is done at intervals (e.g. weekly) instead of being present in surplus as was required with the 'first-generation' anticoagulants. This method of baiting restricts the availability of the poison, whilst allowing animals that had not fed initially the opportunity to feed during a second or subsequent baiting, since dominant individuals that may have restricted their access to the bait would now be dead.

The euphoria attending the release of the 'second-generation' anticoagulants was, however, partly abated by the discovery of cross-resistance to difenacoum in certain areas of England (Redfern and Gill, 1978). Since then, rat infestations on farms in the Hampshire area in particular have become increasingly difficult to control with either difenacoum or bromadiolone.

Rodent bait formulations

In the past the majority of rodenticide baits were presented as loose grain or particulate formulations. The attractiveness of such formulations to non-target bird species in particular has resulted in renewed interest in solid wax blocks, which were developed many years ago to protect bait used in damp conditions such as sewers or in some tropical countries. The sizes of the wax blocks in current use make them ideal for baiting rat holes and burrows and first-hand experience indicates that they present far fewer risks of primary toxicity to bird species than loose grain baits. However, they may also present increased risks to some other non-target species (through carriage by rats or careless use), particularly to domestic and farm dogs which have been known to find the wax blocks attractive to chew on.

The current tendency to supply ready-prepared baits is enabling the commercial formulator to prepare more effective baits by incorporating 'stickers', mould inhibitors, insecticides, anti-oxidants, novel flavours, aromas and taste enhancers, and the addition of dyes and emetics can reduce non-target hazard. Several of these additives have been listed by Marsh (1986).

FUTURE ASPECTS OF RODENT CONTROL

Rodent control in the future does not necessarily imply the wholesale abandonment of currently used rodenticides of proven efficacy in favour of more toxic compounds that are potentially more hazardous to the environment. Rather it should also involve the more imaginative employment of rodenticides and other chemicals, including synthetic pheromones and chemosterilants, through the better utilization of our knowledge of rodent biology.

The following ideas are worthy of further consideration.

Modified baiting techniques

Anticoagulant rodenticides are conventionally applied without prebaiting and this includes the 'second generation' chemicals which are in effect single-dose poisons. It is assumed, from the experience with the 'first generation' anticoagulants, that they act as their own prebait. However, rats in particular and other rodents exhibit neophobic or xenophobic responses to new objects including bait containers and poison baits, and their detection of, and subsequent shyness to anticoagulant baits have been reported (Bowerman and Brooks, 1972; Krishnamurthy, Ramasivan and Singh, 1971). Drug-induced aversion in rats was demonstrated by Elkins (1973) to be a graded response that took up to three weeks to block completely. Such aversion is more likely to occur after sublethal intakes of the more toxic single-dose poisons, including the 'second generation' anticoagulants. However, such aversion or bait shyness may be overcome by prior familiarization with the untreated bait base. Preliminary results from field experiments currently in progress suggest that prebaiting for two weeks can markedly improve anticoagulant rodenticide bait intake, as well as the efficacy of 'second generation' rodenticide treatments, without increasing the overall duration of the treatments (Lazarus, unpublished data). This could in turn have implications in helping to deal with the spread of resistance to 'second generation' anticoagulants, which might in part be due to inadequate baiting methods resulting in sublethal dosing and the survival of rats with low levels of resistance to the anticoagulants.

The further development of selective rodenticides

It is interesting to note that two of the chemicals showing selective toxicity to rats, norbormide and alpha-chlorohydrin, both affect the blood circulatory system. Further research into the reasons for their selectivity to *Rattus* spp., particularly *R. norvegicus*, could result in the eventual production of more selective rodenticides that are less harmful to non-target species. The unpalatability of norbormide and other selective chemicals, and the bait shyness caused, may be overcome by improved bait micro-encapsulation techniques. Alternatively, it would be advantageous to incorporate these selective chemicals in improved contact rodenticide formulations that rely, not on food intake but on grooming behaviour for their ingestion. Such formulations, coupled with improved techniques for the strategic placement of contact rodenticides in areas frequented by rodents, could provide a safe and efficient alternative method to baiting for controlling troublesome rat and mouse infestations.

CO-ORDINATED CONTROL PROGRAMMES

Resistance to the anticoagulant rodenticides will continue to be a problem and increase so long as resistant rodents are selected by the continued inefficient and indiscriminate use of anticoagulant baits. In rodent populations containing a resistance gene, poorly conducted rodenticide treatments are less likely to kill animals having this gene. Such a situation might already exist in Hampshire and the neighbouring counties in the south of England where it is becoming increasingly difficult to control rat infestations on agricultural land using the currently available anti-coagulant rodenticides. Should the problems of anticoagulant resistance in these vulnerable areas continue or become worse, then it may be necessary for all concerned to agree to an integrated rodent control strategy, perhaps involving some temporary cessation of anticoagulant use whilst placing greater reliance on the use of non-anticoagulant control methods. The development of an integrated strategy utilizing co-ordinated area programmes such as evidenced in the recently successful project for controlling rats in Kuwait (Al Sanei, Zagloul and Balba, 1984) but with perhaps less reliance on anticoagulant baits could result not only in significant reductions in the levels of anticoagulant resistance but also in the wholesale elimination of rat infestation in the treated areas. However, this will only be possible if the true scale of the problem is acknowledged by pesticide manufacturers, servicing companies, local authorities, farmers and their advisers.

A more professional and forward-looking approach needs to be adopted by the pest-control industry since it must now be recognized that inefficient control measures can often seriously exasperate an already problematical situation. Greater resources must be directed into research areas aimed at improving the effectiveness of currently used rodenticides and at targeting them more specifically against the infestations to be controlled. Promising novel rodent control methods, especially those with the potential for reducing the risks of killing non-target species, or of reducing the environmental hazards presented by currently used chemicals, must be fully exploited. Rodent pest species are a problem partly because of the rapid rate at which they breed. It would seem more logical therefore to employ measures to prevent them breeding than to try to kill them after they have grown and become established in an infestation and are causing damage. The efficacy of long-lasting reproduction inhibition would be even more effective if chemosterilants sterlizing both sexes (e.g. long-acting cyclopentyl oestrogens etc.) were employed because for the first time we would have the opportunity to affect animals exhibiting 'behavioural resistance' (i.e. those not being attracted to or not taking sufficient quantities of poison bait or succumbing to other means of control), because it takes two fertile animals mating to reproduce successfully. Other advantages of this approach could include increased territorial behaviour by the gradually dwindling sterile population. This would rule out any post-treatment population explosion or 'rebound effect', as can occur after conventional rodenticide treatments. There is also the consideration that any non-target victims of accidental exposure would remain alive for remedial treatment.

Most rodent control measures rely, and will continue to rely on the target animals voluntarily ingesting a harmful chemical. In these circumstances it is very likely that some rats and other rodents prone to neophobia could react unfavourably, often by rejecting the bait in which the harmful chemical is presented. Prebaiting is the method used to overcome neophobia in rodents exposed to traditional acute or single-dose rodenticides. It is not employed for anticoagulants despite the fact that most 'second generation' anticoagulants can act as single dose poisons at the concentrations used in bait. The preliminary trials undertaken suggest that prebaiting can maximize the intake of 'second generation' anticoagulants making them more effective against the target species while reducing by two-thirds the duration of poison baiting. Additional benefits of this method of bait application include the opportunity to monitor by tracking census, etc., the non-target animals at risk from particular bait points and to take exclusion measures before it is too late. The indications are that all this could be accomplished without significantly increasing treatment times or manpower requirements.

REFERENCES

Al Sanei, K., Zagloul, T. and Balba, M.C. (1984) Organisation of the rodent control project in Kuwait in *The Organisation and Practice of Vertebrate Pest Control* (ed. A.C. Dubock), ICI, London, pp. 143–50.

Baker, B. (1983) About glue traps: Some are better than others. *Pest Control,* **51**, 5.

Balser, D.S. (1964) Antifertility agents in vertebrate pest control. *Proc. 2nd Vertebrate Pest Conference*. Anaheim, Calif. pp. 133–7.

Bellier, L. (1973) Rodent pests of West Africa. *Span,* **16**, 28–30.

Berry, R.J. and Jakobson, M.E. (1975) Adaptation and adaptability in wild living house mice (*Mus musculus*). *J. Zool. Lond.,* **176**, 391–402.

Bowerman, A.M. and Brooks, J.E. (1972) Bait shyness. What makes a palatable warfarin? *Pest Control,* **40**, 22, 28, 29.

Boyle, C.H. (1960) Case of apparant resistance of *Rattus norvegicus* Berkenhout to anticoagulant poisons. *Nature, Lond.,* **188**, 517.

Crabtree, D.G. (1950) Raticidal potentialities of WARF-42. *SOAP Sanit. Chem.,* **26**, 131–5.

Davis, D.E. (1961) Principles for population control by gametocides. *Trans. N. Am. Wildl. Conf.,* **26**, 160–7.

Drummond, D.C. (1971) Rodents and biodeterioration. *Int. Biodetn. Bull.*, 7, 73–9.
Drummond, D.C. (1974) Vertebrate pest management in the United Kingdom in *Proc. Regional Training Seminar on Field Rat Control and Research.* RCR/195.
Drummond, D.C. (1980) Pottery rodent traps – a provisional list. *Museum Ethnographers' Group Newsletter* No. 9 March 1980, pp. 14–15.
Drummond, D.C. and Bentley, E.W. (1967) The resistance of rodents to warfarin in England and Wales. *EPPO Publications Series A*, no. 41, 57–67.
Dubock, A.C. (1980) The development and practical use of the novel anticoagulant rodenticide brodifacoum. *Plant Prot. Bull. (Taiwan ROC)* **22**, 223–38.
Dubock, A.C. (1982) Pulsed baiting – a new technique for high potency, slow-acting rodenticides. *Proceedings of 10th Vertebrate Pest Conference*, Monterey, Calif., 23–25 February 1982, pp. 123–6.
Elkins, R.L. (1973) Attenuation of drug induced bait shyness to a palatable solution as an increasing function of its availability prior to conditioning. *Behav. Biol.*, 9, 221–6.
Ericsson, R.J. (1982) Alpha-chlorohydrin (Epibloc): a toxicant-sterilant is an alternative in rodent control. *Proceedings 10th Vertebrate Pest Conference*, Monterey, California, 23–25 February, 1982, pp. 6–9.
Franco, E. and Stazzi, A. (1970) Present aspects of the biological control of rodents. *Boll. 1st Sieroter. Milan*, **49**, 245–67.
Frantz, S.C. and Padula, C.M. (1983) A laboratory test method for evaluating the efficacy of glue boards for trapping mice in *Vertebrate Pest Control and Management Methods, ASTM STP* 817 (ed. D.E. Kaukeinen), American Society for Testing Materials, Philadelphia, pp. 209–25.
Freeman, R.B. (1954) Properties of the poisons used in rodent control in *Control of Rats and Mice* (ed. D. Chitty) vol. 1, Clarendon Press, Oxford, **2**, 214, 220, 244–50.
Garforth, B. and Johnson, R.A. (1987) Performance and safety of the new rodenticide flocoumafen in *Stored Products Pest Control* (ed. T.J. Lawson), *British Crop Protection Monograph* 37, BCPC Publications, Thornton Heath, pp. 105–123.
Girish, G.K. and Arora, K.K. (1976) Studies on rodents and their control. Part XIII Chemosterilant techniques for rodent control [Review]. *Bull. Grain Technol.*, **14**, 219–28.
Gratz, N.G. (1973) A critical review of currently used single-dose rodenticides. *Bull. World Health Organization*, **48**, 469–77.
Gratz, N.G. (1984) The global public health importance of rodents. *Proc. Conf. on the Organization and Practice of Vertebrate Pest Control* (ed. A.C. Dubock), ICI, London, pp. 413–35.
Greaves, J.H. (1971) Rodenticides. *Rep. Progr. Appl. Chem.*, **56**, 465–73.
Greaves, J.H., Shepherd, D.S. and Gill, J.E. (1982) An investigation of difenacoum resistance in Norway rat populations in Hampshire. *Ann. Appl. Biol.*, **100**, 581–7.
Howard, W.E. (1968) Controlling rodents biologically and with sound. *Proc. Rodents as Factors in Disease and Economic Loss*. Honolulu, Hawaii, June 17–27, pp. 220–41.
Jackson, W.B. (1987) Current management strategies for commensal rodents. *Current Mammalogy*, **1**, 495–512.
Kaufman, G.E. and Michaelson, S.M. (1974) Critical review of the biological effects of electric and magnetic fields in *Biological and Clinical Effects of Low-frequency Magnetic and Electric Fields* (eds J.G. Llaurado *et al.*) C.C. Thomas, Illinois, pp. 49–61.

Kendle, K.E., Lazarus, A.B., Rowe, F.P., Telford, J.M. and Vallance, D.K. (1973) Sterilization of rodent and other pests using a synthetic oestrogen. *Nature*, **244**, 105–8.
Knipling, E.F. (1959) Sterile male method of population control. *Science*, **130**, 902–4.
Krishnamurthy, K., Ramasivan, T. and Singh, D.P. (1971) Studies on rodents and their control Part VII: Effect of impurities in warfarin on its acceptability and mortality to black rats (*Rattus rattus*). *Bull. Grain Technol.*, **9**, 252–6.
Lazarus, A.B. and Rowe, F.P. (1982) Reproduction in an island population of Norway rats, *Rattus norvegicus* (Berkenhout), treated with an oestrogenic steroid. *Agro-Ecosystems*, **8**, 59–67.
Linhart, S.B. and Enders, R.K. (1964) Some effects of diethylstilbestrol on reproduction in captive red foxes. *J. Wildl. Mgmt*, **28**, 358–63.
Lund, M. (1982) Rats resistant to anticoagulants. *Danish Pest Infestation Laboratory Annual Report* 1981, pp. 60–1.
McDonald, M. (1980) Population control of feral cats using megestrol acetate. *Vet. Rec.*, **106**, 129.
Mallis, A. (1982) *Handbook of Pest Control*, 6th edn. Franzak and Foster, Cleveland, Ohio.
Marsdin, R. (1975) Mice infestation: Chewing of electric cables. *Environ. Health*, **83**, 345.
Marsh, R.E. (1983) Tracking Powders. Providing effective rodent control. *Pest Control*, **50**, 42–8.
Marsh, R.E. (1986) Principles and techniques of formulating effective rodent baits – present and future. *Proceedings of the Second Symposium on Recent Advances in Rodent Control*, Kuwait, pp. 211–23.
Marsh, R.E. and Howard, W.E. (1969) Evaluation of mestranol as a reproductive inhibitor of Norway rats on garbage dumps. *J. Wildl. Mgmt*, **33**, 133–8.
Meehan, A.P. (1984) *Rats and Mice: Their Biology and Control*. Rentokil Ltd., East Grinstead, West Sussex.
Ophof, A.J. and Langeveld, D.W. (1969) Warfarin-resistance in the Netherlands. *Schrreihe Ver. Wass.-Boden-Lufthyg. Berlin-Dahlen*, **32**, 39–47. (WHO/VBC/**68**, 109).
Plant, R. (1983) Chemical rabbit control. *Big Farm Management* (April) 22.
Redfern, R. and Gill, J.E. (1978) The development and use of a test to identify resistance to the anticoagulant difenacoum in the Norway rat (*Rattus norvegicus*). *J. Hyg., Cambridge*, **81**, 427–31.
Rowe, F.P. and Lazarus, A.B. (1974) The effects of an oestrogenic steroid on the reproduction of wild rats, *Rattus norvegicus* (Berk.), *Agro-Ecosystems*, **1**, 57–68.
Smith, R.H. and Greaves, J.H. (1986) Resistance to anticoagulant rodenticides: The problem and its management in *4th International Conference on Stored-product Protection*. Tel Aviv, Israel.
Wilson, E.J. (1960) Ultra-violet lamps and rodent control. *Parasitica*, **16**, 119–25.
Wilson, E.J. (1972) The assessment of rat damage. *South Pacific Commission Information Document* no. 28, pp. 1–16.
Wodzicki, K. (1973) Prospects for biological control of rodent populations. *Bull. Wld Health Org.*, **48**, 461–7.

Seven

Moles as pests

R.D. STONE

SUMMARY

On the basis of their life style, subterranean mammals may be divided into two main groups: obligate fossorial species which are totally dependent on the tunnel systems which they create to obtain food, to provide shelter and protection; and incidental species, which utilize the permanent or temporary series of tunnels created by others, exploiting the resources therein. Of the several hundred species belonging to this former group only a small proportion is considered to pose any serious economic problems and, in many of these cases, it is the incidental species which may be to blame for the damage caused. All of the obligate fossorial species are confined to two orders: Rodentia and Insectivora, and although they display a diverse array of morphological and ecological adaptations, there is one particular trait which they all have in common – the evacuation of earth onto the surface in the process of tunnel formation. Thus, largely because of damage to agricultural and recreational sites, pest status has been assigned to these species.

This chapter describes the ecology of one of these obligate fossorial species, the European mole (*Talpa europaea*). Although all fossorial moles are generally considered as pests, the emphasis for control has always been on the single European species. Because of this, the chapter will concentrate on the reasons why moles are classified as pests and it will evaluate the control techniques, both past and present, as well as the need to implement these control measures. These, in turn, will be compared with those techniques applied to control other fossorial species.

INTRODUCTION

Many small mammals live in close association with man (for example, the house mouse *Mus musculus*, and house shrew *Suncus murinus*), but some species, in particular the rodents, live less harmoniously with man than others. Such species are generally considered pests because they interfere with man's agricultural or social activities. For the purpose of this chapter, a pest species may be defined as 'a troublesome, annoying or destructive person, animal or thing' (Oxford Dictionary, 1982). In recent years the tension between man and 'pest' has increased dramatically in the field of agriculture, where a continuing trend towards extensive cultivation of large areas of land with a single crop species enables certain species to attain higher densities than they would otherwise reach with natural vegetation.

This chapter concentrates on a fossorial species, the European mole (*Talpa europaea*), and its interactions with man. In particular it will review the reasons why moles are considered to be a pest species, what steps have been taken to control these animals in the past, and whether these have been justified and effective. In conclusion, the chapter will make some suggestions as to alternatives to current methods of control. (For a general introduction to the ecology of this species, see Godfrey and Crowcroft, 1960, and Stone, 1986a). While the main emphasis will be on the European species, *T. europaea*, wherever possible reference will also be made to comparative work on North American moles and occasionally to other fossorial species, both rodents and insectivores. In general, however, little consideration will be given to fossorial voles (genus *Arvicola* and *Pitymys*) or to those voles which may live predominantly below ground but which feed on epigeal vegetation (genus *Microtus*). Suffice it to say that damage caused by subterranean voles is often acute, since the root systems of plants which have been destroyed do not recover.

WHY ARE FOSSORIAL SPECIES CONSIDERED TO BE PESTS?

There are three main reasons why fossorial species in general, and moles in particular, are considered as pests. First, they may be the cause of minor nuisance involving small economic loss, for example, when moles tunnel across lawns, bowling greens, golf courses or cricket greens, throwing up molehills as they dig. These are either considered to be aesthetically unattractive, or are simply an inconvenience. Secondly, and increasing somewhat in the seriousness of the problem, the tunnels and spoil heaps of fossorial species often provide the means for invasion and colonization by other species, both plant and animal; these later species may then prove to be far more destructive to crops than were the original inhabitants. Often this colonization of existing tunnels by an invading species leads to the expulsion of the original inhabitants. Finally, and of much greater economic importance is the direct damage which fossorial species cause to agriculture. This aspect will be considered in more detail later.

On the other side of the coin, some fossorial species were once thought to be beneficial to agriculture by improving the aeration and drainage of the soil (*Talpa europaea*, *Cryptotis hottentotus*), destroying harmful soil insects (*T. europaea*, *Scalopus aquaticus*, *Scapanus orarius*), and circulating soil minerals (Abaturov, 1972; Demela, 1950). However, these benefits have now been supplanted by modern agricultural practices and moles are now considered as pests by the vast majority of people with an involvement in agriculture. During the last century and, indeed, up to the 1940s, the European mole (*T. europaea*) was commercially exploited for its valuable pelage which was used in the clothing industry. However, with the introduction of modern-day textiles, this once-thriving industry ceased to flourish.

WHAT DAMAGE DO FOSSORIAL SPECIES CAUSE?

The main forms of damage caused to agriculture by moles include: (1) uprooting of seedlings as the animal raises the roof of the tunnel in tunnel construction, thereby either killing them directly, or depriving the rooting system of soil and water. This disturbance may also lead to exposure of roots to frost. (2) As already noted, secondary use is made of existing mole tunnels by many other animal species (e.g. *Pitymys* spp., *Microtus* spp.), which eat the exposed roots and bulbs. (3) Excavated soil from molehills can cover and thus kill an estimated 10% of a sward within the territory of a single animal (Schaefer, 1979). (4) Molehills may cause considerable damage to farm machinery at harvesting time, through contact with cutting blades and the excavated earth, or stones which moles may have pushed to the surface. In certain areas molehills may also become colonized by ants, which effectively hardens the exterior of the molehill and may therefore cause considerable damage to machinery. (5) Molehills may, in some areas, lead to soil erosion. (6) Molehills are often the first areas in fields to be colonized by weed

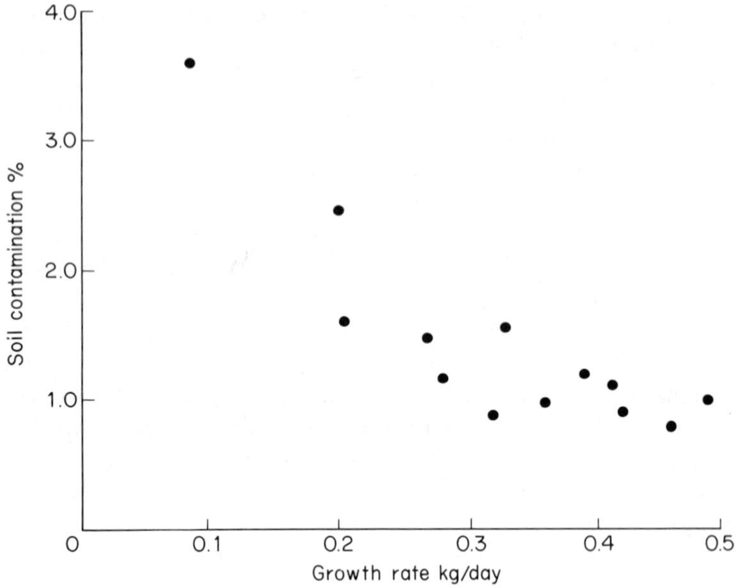

Figure 7.1 The performance of yearling beef cattle conducted over a three-year period of varying levels of soil contamination.

species, which may then invade the pasture and further dilute the sward. Thistles (*Cirsium* spp.) are particularly adept at colonizing molehills and, in one experiment, Ford (1935) reported that the number of thistles per square yard on mole-infested ground was 6.9, in comparison to 0.7 in an adjoining region free of moles (see also Jalloq, 1975). Finally, (7), hay and particularly silage crops, the latter depending on a fermentation process for preservation, may become contaminated with soil. This, in turn, may lead to the partial or complete loss of the stored crop or to disease and loss of quality within a herd. In a three-year study conducted by the Ministry of Agriculture, Fisheries and Food (MAFF) in the UK, the performance of yearling beef cattle was found to be markedly poorer where soil contamination rose above 2.0% (Figure 7.1). It should be stressed, however, that soil derived from molehills is by no means the only source of contamination within a crop. The type of mower, the cutting height, the harvesting machinery and the weather conditions at the time of cutting may all contribute to the amount of soil taken in with a crop. For example, at open pit silos a considerable amount of mud and earth may be mixed in with the silage each time a tractor adds more grass to the pit. Contamination from farmyard manure is another common and serious cause of poor fermentation and preservation in silage.

WHAT FACTORS DETERMINE MOLE DISTRIBUTION?

For *Talpa europaea*, and some other fossorial species, the abundance of animals appears to be directly related to the habitat (Funmilayo, 1970). Moles are likely to be limited in their distribution by three main environmental factors: first, soil topography, i.e. the ability of the soil, or litter layer, to sustain an adequate tunnel system. Moles produce fewer hills in areas where the soil is dry and dense, probably because the soil becomes more difficult to displace when it is heavier and when it is drier, and/or because burrows in dense ground stay open longer and require less maintenance. The second factor influencing or potentially influencing mole distribution is the ability of the soil to produce sufficient invertebrate food to sustain a viable population of moles, a factor related to the percentage of organic matter content. Indeed, earthworm populations vary considerably both in total numbers and in species composition within small areas of a field, a factor which has been related to local soil conditions such as soil moisture and soil temperature. Finally, the pH level of the soil may also influence mole distribution and density. For the coast mole, *Scapanus orarius*, at least, correlations between molehill numbers and soil mineral levels showed that the number of molehills produced was significantly positively correlated with pH (Schaefer, 1979).

The principal dietary component of all mole species which have been studied to date, is earthworms. However, the percentage weight of earthworms in the diet may vary for a single species within different localities (Funmilayo, 1970). It is only under extreme conditions where earthworms are scarce that insect larvae or even ants will be consumed in large quantities. It appears that there is no correlation between the number of molehills and earthworm species (Funmilayo, 1970; Schaefer, 1979; Stone, unpublished).

WHY ARE CONTROL METHODS NECESSARY?

Obligate fossorial species such as moles have no natural enemies which consistently prey upon them. Certain species such as owls (*Strix aluco*, Southern, 1954), buzzards (*Buteo buteo*, Skoczen, 1962), herons (*Ardea ardea*, M. Marquiss, personal communication), and certain carnivores will prey on moles only if the opportunity presents itself. The single most important stage of mortality for moles, and indeed all fossorial species, is during the dispersal phase, which occurs primarily within the juvenile cohort between May and July, depending on latitude. At this time of year, juveniles are expelled from the maternal burrow system and forced to move above ground in search of new territories. Apart from this single factor, moles seem to be limited only by their food supply and the area of suitable habitat. Severe frosts do not appear to reduce mole numbers – moles generally excavate deeper tunnels to locate prey.

Flooding from exceptionally heavy rain and spring run-off causes some mortality (Schaefer, 1979), but this effect is usually localized and of little significance since moles are known to be capable swimmers (Godfrey and Crowcroft, 1960; Mellanby, 1971; Stone, personal observation). However, it is likely that under prolonged conditions of flooding moles may die from starvation or migrate because of earthworm mortality.

POPULATION DENSITY ESTIMATES

Effective control measures on any species rely heavily on indices of population levels. However, arriving at such estimates is an especially acute problem for fossorial mammals, since these are rarely observed. In the past, various techniques have been designed in an attempt to overcome this problem. One of the most commonly used techniques has been the trapping out of all animals within a certain delineated area. Such an approach makes the assumption that all animals are likely to be trapped to the same degree and that all animals within that area were trapped. However, unless the trapping operation is continued over a sufficiently long period of time, neither of these assumptions holds true. The reasons for this are: first, within a given species not all animals can be trapped equally. For the majority of mammals, males are more frequently trapped than females, especially during the breeding season (see also Gosling and Baker, Chaper 14). This is particularly evident in *T. europaea*. Another problem arises from operator control – the way in which traps are handled and set (Rudge, 1963; Stone, 1986b). This latter point is especially valid when trapping fossorial mammals since most species appear to have an acute spatial memory and are able to detect disturbances to their tunnel system caused by trap setting. Trap odour is also an important consideration: new traps are usually avoided by all animals, whereas a seasoned trap which has previously held a female mole is more likely than a clean trap to attract a male. Traps which are detected are usually plugged with earth and the area avoided thereafter for several days. This is illustrated in Table 7.1, which shows the extent of trap blockings by *T. europaea* (Stone, 1986b). Unless taken into consideration, the number of misses, such as those shown in Table 7.1, will have an effect on the population estimate. Finally, the social organization of a particular species should also be taken into account when estimating population density since the number of potential captures with a single

Table 7.1 A comparison of the trapping success of models of mole traps in relation to the number of days set

Trap	Days set	Captures	Misses		Untouched
			Sprung	Avoided	
Duffus	540	421	24	15	80
Frisian	264	113	31	45	75
Pincher	360	203	53	48	56
Total all traps	1164	737	108	108	211

Data from north-east Scotland (Stone, 1986b).

trap at one site will depend on whether the species is solitary or social, territorial or not (Stone, 1986b).

Several attempts to quantify the number of individuals of a particular fossorial species have relied on indices based on molehill formation. For some species, such as *Spalax ehrenbergi*, a mole rat, each solitary individual living underground is identifiable by mound aggregations of 5–30 mounds, comprising its territorial domain (Nevo, 1961). For this species, the individual territories can be easily scored at the beginning of the breeding season. Unfortunately, it is uncertain whether this holds true for all mole rat species, and it is not reliable for insectivorous fossorial species, since the extent of tunnel construction is dependent on such factors as soil topography and prey biomass. Nonetheless, many authors believe this to be a most useful technique, although as Goszczynski (1983) has shown, the digging activities of individual animals vary considerably throughout the year. In addition spring activities of moles are more noticeable among males than females, so that there is a high possibility of estimate bias in favour of males at this time of year (Stone, 1986b). Another factor which is rarely taken into consideration is the breakdown of molehills by climatic effects (Jonca, 1972), animals or other factors.

Where absolute estimates of animal numbers are impracticable, many workers rely on various indirect indices of abundance. For moles, population indices may be derived from physically destroying existing molehills, surface tunnels and surface openings (see Mead-Briggs and Woods, 1973). The animals' activities, in terms of molehill production, are then recorded and an index of activity calculated.

Mole population densities never reach the levels of many other fossorial species. From a three-year sample of trapping in different habitats in north-east Scotland, mole density was always found to be around five animals per hectare (Stone, 1986b). By contrast, *Arvicola terrestris sherman* may reach densities of over 5000 animals per hectare

(A. Meylan, personal communication), whereas population densities for mole rats average 16 animals per hectare for solitary species, and up to 50 animals per 100 m^2 for a social species. Values for *T. europaea* seem minimal by comparison.

MOLE CONTROL

The practice of mole control has undergone a steady change since the beginning of this century; from the once highly skilled, professional art, whereby a molecatcher would set a single noose snare with a willow sapling for each animal, it has shifted to the semi-skilled laying of a wide array of traps, and to current techniques involving the laying of poisoned baits or gas-releasing pellets.

Trapping

Trapping is one of the safest and most effective means of getting rid of pest species and, as one of the oldest forms of killing animals, many models have been designed to fulfil this function. Thus, whereas earlier traps, such as the harpoon trap, were large, cumbersome and awkward to set, modern-day models are light and simple to use. The two most popular types of mole trap currently available on the market are the scissors or pincer trap and the 'Duffus' or half-barrel trap. The efficiency of these two traps was tested against that of a live-capture trap, the Frisian trap as designed by Haeck (1961); results of these trials are presented in Table 7.2. It is apparent that, in terms of catching animals, the Duffus model is the most efficient of the three, although data from Rudge (1963) suggest the contrary.

The main, single objection raised against the use of traps for widespread mole control has been that the technique requires skilled labour

Table 7.2 A comparison of the efficacy of three mole traps

	Duffus	Frisian	Pincher
Percentage of traps sprung (no catch) and avoided in relation to settings	7	29	28
Percentage of captures in relation to settings	78	31	56
Percentage of traps sprung (captures and misses) in relation to settings	82	54	71

Data from north-east Scotland (Stone, 1986b).

labour and is extremely labour-intensive, since traps must be frequently revisited and, on occasion, re-set. However, it should be remembered that trapping does yield carcass material which may be extremely valuable for laboratory and biological analyses and, as we shall see later, trapping does not pose any damaging secondary effects to the environment. Trapping is also, arguably, the most humane method of mole control.

Poisonous compounds

Strychnine. The success of strychnine in mole control programmes is almost unfortunate (Neville, 1985). An earthworm baited with strychnine powder inserted into a mole run, will usually kill the mole. While trapping can be just as effective for small-scale infestations, with large-scale agricultural mole problems it becomes too time-consuming to set the required number of traps and, until recently, there were no methods other than strychnine baits which provided control as cheaply, quickly and effectively.

Strychnine is an alkaloid compound found in the seeds of the plant *Strychnos nux vomica* and other related species (Clarke and Clarke, 1967). Strychnine can be isolated from the seeds and, in the form of hydrochloride or sulphate salt, is an acute poison. Examples of LD_{50} values following oral administration are given in Table 7.3. No precise LD_{50} value is available for moles, but it is thought to lie between that of the rat and mouse, at about 0.9 mg/kg (Neville, 1985). The poison is presented to moles in a baited earthworm. The dose used is obviously sufficient to kill the mole but the amount ingested varies with the size of the worm and the thoroughness with which the mole cleans the worm prior to ingestion. When feeding, the worm is held between the mole's front paws and eaten head first in a series of jerky movements, by which the mole pulls the worm between its claws, thereby removing excess grit from the outside of the worm and also forcing out some grit from the worm's digestive tract. It is believed that the usual time between ingestion

Table 7.3 Fifty per cent lethal dose levels of orally administered strychnine. Values given are mg strychnine per kg body weight

Cattle	1.5
Rat	1.0
Mouse	0.8
Cat/Dog	0.75

Data after Neville (1985).

of strychnine and the death of the mole is between 15 minutes and two hours, depending on those factors already mentioned. Although observable symptoms may not be exhibited for some time after ingestion (Neville, 1985), once symptoms commence, death follows 15–30 minutes later.

There are three main disadvantages and items of concern governing the use of strychnine. First, the technique is inhumane – animals given a standard dose of strychnine take an agonizingly slow time to die. Secondly, strychnine presents a serious risk of secondary poisoning both to man and other vertebrate species. Finally, operational costs are high.

Alphachoralose. For many years the development of a suitable alternative to strychnine has been sought and, in 1978, the Universities Federation of Animal Welfare (UFAW) investigated whether worms baited with alphachloralose could be used to control moles. Alphachloralose is a human narcotic agent already in restricted use for the control of several pest species of birds and rodents. With this technique, earthworms are baited with alphachloralose either by dusting with powder or by immersion in a thick paste; in either case, worms are narcotized by the alphachloralose and are therefore unable to burrow.

In a series of trials carried out by UFAW, active mole workings were identified by the presence of fresh molehills, five baited worms laid on the tunnel floor and the opening resealed (Neville, 1985). The tunnels were later reopened to record the number of baits taken or disturbed by moles. In a few cases, all five baits were consumed and fresh molehills were not produced at any workings where at least three baits treated either with paste or powder were taken. Moles were presumed to have been killed in these instances and indeed two specimens were recovered from under the turf. However, in all trials, only about half the number of active moles sampled the baits and most rejected them after consuming less than half a worm. No reduction in molehills was achieved in these instances.

In a comparative trial, also carried out by UFAW, active workings treated with single worms baited with strychnine became inactive and no further hills were produced for several weeks. Single alphachloralose baits applied to similar infestation caused an approximate 50% decrease in molehill production for 24 hours, but the original level of activity returned after 48 hours, suggesting that they were narcotized by the alphachloralose, but did not receive a lethal dose.

These field investigations have shown that alphachloralose-baited worms can achieve a temporary reduction in molehill production but are only fatal to the mole if consumed in sufficient quantity. However, in these trials, the level of control achieved did not compare favourably with strychnine and would not be sufficient for alphachloralose, presented as described, to be recommended as a replacement agent. It is undoubtedly unpalatable to moles who often apparently detected and rejected the baits without tasting them. To be able to utilize the humane killing potential of alphachloralose, a method must be found to disguise its smell and taste and it must be presented in a more palatable form.

Sodium fluoracetate. Trials have also been conducted by MAFF with sodium fluoracetate (Pest Infestation Control 1974–1976). Using treated earthworm baits, each worm carrying approximately 15 mg of poison, laid at a density of 25/ha, a reduction of 44% in mole activity was assessed by the method of Mead-Briggs and Woods (1973). In a follow-up trial at the same site, using bait set at a density of 50/ha, an additional drop of 9% was indicated – a 53% reduction overall. In this series of trials it appeared likely that some bait rejection had occurred. In addition, two moles were found dead on the surface following the poisoning treatment, a fact that could indicate a secondary poisoning hazard. The comparatively low kill achieved in this trial and the possibility of secondary poisoning suggested that sodium fluoracetate was not a suitable alternative to strychnine.

Other chemical control methods

Extensive trials have also been made in controlling moles with a variety of techniques in Denmark. One positive approach to this work was the distribution of a questionnaire in 1976 on the subject of mole control within the Danish municipalities (Lund and Lodal, 1976). Of the 275 Danish municipalities which were contacted, 261 replied, of which 27 stated that they did not control moles whereas the remaining 234 had a control programme. At this time, the most commonly used method for control was that of baiting with earthworms impregnated with a 2% thallium sulphate solution alone (75.2% of the respondents) or combined with other methods such as the use of scissor traps.

The evaluation of the effectiveness of the mole control programme was found to be very subjective, as some municipalities evaluated their results in relation to their expenses, whereas others evaluated the effectiveness in accordance with the number of complaints lodged about the control techniques. Yet, regardless of the basis of this evaluation, it was apparent that 137 (58.5%) of the Danish municipalities were satisfied with the effectiveness of the control programme. Thallium sulphate has now been banned from use since 1979 as it was believed to involve a risk of poisoning non-target animals and a risk of environmental pollution (Lodal, 1980).

In an attempt to find a suitable alternative to thallium sulphate, field trials were begun using carbon dioxide pellets and gas as a control for moles (Lodal, 1980). Mole activity in treated and control areas was assessed by levelling all molehills and opening prod holes in the tunnel system. After a period of two days, the number of new heaps and the number of sealed holes were recorded. However, the results showed no marked reduction in mole activity within the treated area, and further trials are not planned.

Finally, and once again in Denmark, phosphine-generating pellets have been given extensive trials as a method of mole control (Lund, 1975; Lodal, 1981). With this technique, two or three pellets (each weighing 0.6 g), are inserted into every 2–3 m of active tunnel system. A soil temperature of 6–8°C or more is considered optimal. The pellets contain 57% aluminium phosphide and, when in contact with moisture, each pellet develops about 0.2 g phosphine gas. Results obtained from this technique have proved it to be extremely successful. More recently, Lodal (1984) has observed that the effective dosage level of phosphine may be reduced to as little as one (0.6 g) pellet for every 2–3 m of tunnel system, although higher dose rates are necessary for sandy soils.

Outside Europe, poisoned baits, poisoned gases, soil fumigants, pesticides, and land husbandry techniques have all been largely unsuccessful in controlling the coast mole (*Scapanus orarius*), although land husbandry was believed to be the most effective application (Glendenning, 1959). Kuhn (1970) has successfully used dry pelleted baits containing 1% thallium sulphate to control the Townsend mole (*Scapanus townsendii*).

ALTERNATIVES TO KILLING MOLES

Under certain circumstances it may be unnecessary or undesirable to kill moles by trapping or poisoning. This is especially true for private gardens, where the financial costs of employing trained personnel would greatly exceed the damage which a resident mole would be likely to cause. Under such circumstances, many commercial products are available, which, when placed within a mole's tunnel, will release an irritating but non-toxic gas which causes the mole to leave the tunnel system. However, several such applications will be necessary to prevent the animal from returning when the smoke has cleared. Many other humane attempts to get rid of moles, such as ultrasonic emissions, are now being developed, although to date, their efficiency has not been fully tested.

Various other possible control measures have been investigated which do not rely on direct population control, but rather, operate through habitat manipulations. Shilova *et al.* (1971) experimented with insecticides such as aldrin, dieldrin, chlordane and sevin, which act indirectly in mole control by killing the principal food source – earthworms. This solution, however, is not desirable or recommended because of the beneficial effects of earthworms on soil fertility (Evans, 1948; Evans and McGuild, 1947; Edwards and Lofty, 1972) and the negative long-term environmental impacts of such insecticides.

As a form of biological habitat manipulation, Ennik (1967) and Schaefer (1981) both experimented with adding nitrogen-based fertilizers to the soil. The rationale of these experiments was that if nitrogen, or some other acidifying agent, successfully reduced the soil pH level, the activity levels of moles would be reduced, and the number of molehills in an area diminished, indirectly, through a reduction of earthworm numbers and biomass. However, whereas Ennik felt that this technique had much to offer, the more detailed experiments of Schaefer showed that although the soil pH did indeed alter the numbers of earthworms, leading to a slightly reduced activity of moles, at elevated levels of nitrogen application, the increased soil acidity became detrimental to crops.

More recently, a novel means of controlling and restricting the movements of individual moles has been the placing of scented extractions from the preputial glands of moles at the territorial boundaries of neighbouring animals (Gorman and Stone, see Chapter 8). European moles are highly territorial for much of the year and appear to demarcate fixed boundaries with scent secretions which may be replaced daily (Stone, 1986b) during the fixed daily pattern of spatial utilization shown by the majority of moles. Thus, simply by replacing a resident owner's smell with that of secretions from the paired preputial glands, it may be possible to prevent selected neighbouring animals from occupying a vacated territory. This aspect is treated in more detail in the following chapter.

There are many occasions when a little practical application would help minimize the conflict between fossorial species and man. The vast majority of damage caused by fossorial species occurs within a brief period during the calendar year, although it will usually be repeated on a seasonal basis. For temperate region species, such as *Talpa europaea*, peak signs of surface activity are during November and December, coinciding with the onset of cold weather. During such periods, soil invertebrates retreat deeper within the soil level and generally become less active. Thus, with less potential food inhabiting the upper soil strata, moles create new, deeper tunnels, or simply re-excavate those which may have fallen into disrepair during the summer months, so that they may then trap those invertebrates living deeper within the soil. Similarly, during dry periods, new molehills may appear, indicating that the tunnel system is being enlarged for the same reason.

The second major period of activity among temperate dwelling species comes just prior to the breeding season (February to May), when males construct lengthy tunnels, just beneath the soil surface. The object of this exercise is to locate potential mates. By constructing such lengthy radiating tunnels, a single male will be able to make contact with several females. And, finally, following dispersal (July–August), juvenile moles must create their own, new tunnel system if they are unable to locate other unoccupied burrows.

With the exception of moles rebuilding their tunnel systems following mechanical destruction, above ground signs of activity are confined to these periods. Thus, during such periods of 'peak activity', the potentially harmful effects of molehills – which are the primary concern of the general public and farmers alike – could be significantly reduced by the farmers harrowing or rolling their crops (if appropriate, as indeed it is for silage), to flatten the hills, or for those people troubled with moles in their lawn, to simply remove the earth heaps as they are produced. For isolated cases, this is certainly feasible.

In summary, it should be stressed that several measures should be considered before implementing any form of mole control and, in particular, the situation should be carefully assessed before any action is taken. The following points warrant consideration: first, are the animals causing an actual economic loss? If so, is the cost of alleviating this problem likely to surpass the economic loss which might occur if no remedial action were to be taken? If the answer to both these points is positive, then an efficient, humane solution should be sought to alleviate the problem. For fossorial mammals, this generally involves either trapping or poisoning. The former is efficient and generally humane. The latter is more cost effective than trapping and is more easily applied to large areas.

Finally, as research continues to find alternative poisonous compounds to the deadly strychnine, it must be questioned whether or not the costs associated with developing and testing these substances is truly warranted. It is undeniable that certain fossorial species cause considerable economic loss each year, yet for others and, in particular insectivorous species such as the European mole, no truly objective estimates of the damage caused have ever been produced.

ACKNOWLEDGEMENTS

I am very grateful to Dr Rory Putman for his comments on an earlier draft of this manuscript. Special thanks also to Jens Lodal and John Woods for their assistance in providing additional material on control methods.

Support from the Mammal Society to attend this conference is gratefully acknowledged.

REFERENCES

Abaturov, B.C. (1972) The role of burrowing animals in the transfer of mineral substances in the soil. *Pedobiologica*, **12**, 261–6.

Clarke, E.G.C. and Clarke, M.L. (1967) *Veterinary Toxicology*: part 3 *Organic Compounds, Drugs*. Balliere, Tindall and Cassell, London.

Demela, J. (1950) Is the mole in meadows useful or harmful? *Vestnik Ceskoslov. Acad. Zem.*, **24**, 134–6.

Edwards, C.A. and Lofty, J.R. (1972) *Biology of Earthworms*. Chapman and Hall, London.

Ennik, G.C. (1967) Mole activity in relation to pasture management and nitrogen fertilisation. *Neth. J. Agr. Sci.*, **15**, 221–8.

Evans, A.C. (1948) Studies in the relationship between earthworms and soil fertility. I. Some effects of earthworms on soil structure. *Ann. Appl. Biol.*, **35**, 1–13.

Evans, A.C. and McGuild, W.J. (1947) Studies in the relationship between earthworms and soil fertility. II. Biological studies in the field. *Ann. Appl. Biol.*, **34**, 307–30.

Ford, J. (1935) A preliminary investigation on the ecology of the common mole (*Talpa europaea*). *J. Anim. Ecol.*, **4**, 88–9.

Funmilayo, O. (1970) Population studies on the distribution of moles relative to food supply and general habitat. Unpublished PhD Thesis, University of Edinburgh.

Glendenning, R. (1959) Biology and control of the coast mole, *Scapanus orarius*, in British Columbia. *Can. J. Anim. Sci.*, **39**, 34–44.

Godfrey, G.K. and Crowcroft, P. (1960) *The Life of the Mole*. Museum Press.

Goszczynski, J. (1983) Digging activity and estimation of the population density of moles. *Acta Theriol.*, **28**, 328–32.

Haeck, J. (1961) Een val voor het leverd vargen van mollen. *Lutra*, **3**, 24.

Jalloq, M.C. (1975) The invasion of molehills by weeds as a possible factor in the degeneration of reseeded pasture. *J. Appl. Ecol.*, **12**, 643–57.

Jonca, E. (1972) Winter denudation of molehills in mountainous areas. *Acta Theriol.*, **17**, 407–12.

Kuhn, L.W. (1970) Mole control. in *Proceedings 4th Vertebrate Pest Conference* (ed. R.H. Dana), West Sacremento, Cal., pp. 71–6.

Lodal, J. (1980) Field trials with carbon dioxide for mole control. *Danish Pest Infestation Laboratory Annual Report*. pp. 70–1.

Lodal, J. (1981) Field test with phosphine generating pellets. *Danish Pest Infestation Laboratory Annual Report*. pp. 71–2.

Lodal, J. (1984) Field trial with a phosphine generating calciumphosphide preparation. *Danish Pest Infestation Laboratory Annual Report*. pp. 84–5.

Lund, M. (1975) Control of moles: comparison of phosphine generating pellets and thallium sulphate on earthworms. *Danish Pest Infestation Laboratory Annual Report*. pp. 63–5.

Lund, M. and Lodal, J. (1976) Survey of the municipal mole control programme. *Danish Pest Infestation Laboratory Annual Report*.

Mead-Briggs, A.R. and Woods, J.A. (1973) An index of activity to assess the reduction in mole numbers caused by control measures. *J. Appl. Ecol.*, **10**, 837–45.

Mellanby, K. (1971) *The Mole*. The New Naturalist. Collins, London, pp. 159.
Ministry of Agriculture, Fisheries and Food (1977) Moles: Assessment of damage. in *Pest Infestation Control*, pp. 166–71. MAFF Publication.
Neville, P. (1985) Humane control of moles in *Humane Control of Land Mammals and Birds*. Proceedings of a symposium. (ed. D.P. Britt), The Universities Federation for Animal Welfare.
Nevo, E. (1961) Observations on Israeli populations of the mole rat. *Spalax e. ehrenbergi* Nehring 1898. *Mammalia*, **25**, 127–44.
Rudge, J. (1963) A study of mole trapping. *Proc. Zool. Soc. Lond.*, **140**, 330–4.
Schaefer, V.H. (1979) Aspects of habitat selection of the coast mole (*Scapanus orarius*) in British Columbia. Unpublished PhD Thesis, Simon Fraser University, Burnaby, BC.
Schaefer, V.H. (1981) A test of the possible reduction of the digging activity of moles in pasture by increasing soil nitrogen. *Acta Theriol.* **26**(7) 113–17.
Shilova, S.A., Denisova, A.V., Dmitriev, G.A., Voronova, L.D. and Bardier, M.N. (1971) Effect of some insecticides upon the common mole. *Zool. Zh.*, **50**, 886–92.
Skoczen, S. (1958) Tunnel digging by the mole (*T. europaea*). *Acta Theriol.*, **2**, 235–49.
Skoczen, S. (1962) Age structure of skulls of the mole, *Talpa europaea* Linnnaeus 1758 from food of the buzzard (*Buteo buteo* L). *Acta Theriol.*, **6**, 1–9.
Southern, H.N. (1954) Tawny owls and their prey. *Ibis*, **96**, 384–410.
Stone, R.D. (1986a) *Moles*. Anthony Nelson Ltd, Oswestry, England.
Stone, R.D. (1986b) The social ecology of the European mole (*Talpa europaea* L) and the Pyrenean desman (*Galemys pyrenaicus* G): A comparative study. Unpublished PhD Thesis, University of Aberdeen.
UFAW (1983) The use of alphachloralose for the control of moles in *UFAW Report and Accounts*.
UFAW (1985) *Humane Control of Land Mammals and Birds. Proceedings of a symposium* (ed. D.P. Britt), The Universities Federation for Animal Welfare.

Eight

Repelling moles

M.L. GORMAN AND R.D. STONE

SUMMARY

This paper deals with alternative approaches to the control of mole damage. The current methods used in the UK, trapping and poisoning with strychnine, are cruel. A further problem is that strychnine is also used illegally to kill other species, including birds of prey. We have investigated the possibility of using two classes of odorous chemicals to repel moles: social odours produced by carnivores and mole odours.

Weasel anal sac secretions and synthetic 2,2-dimethylthietane added to mole traps significantly reduce their ability to catch moles. We conclude that moles detect and avoid these substances. In a field trial a free-living mole avoided tunnels that had been treated with dimethylthietane.

Free living moles are also repelled by samples of mole preputial gland secretion placed in their tunnels. A preliminary chemical analysis of this secretion is presented. It consists of a complex mixture of compounds whose relative concentrations differ from individual to individual and also between the sexes.

Although moles avoid these two classes of compounds it is unlikely that they will be of use in large-scale mole control.

INTRODUCTION

The purpose of this chapter is to consider alternative approaches to the control of moles, *Talpa europaea*, and the damage that they cause.

In 1982, moles were costing British agriculture very approximately £2 million p.a. (MAFF figures courtesy of Dr A.R. Hardy). This figure includes both monetary loss (largely due to damage to cutting machinery, the loss of grazing and the spoilage of silage) and the cost of control measures. Other damage, for example to gardens and to sports facilities, is more difficult to assess, but undoubtedly it causes genuine nuisance and distress.

The current methods of control in the United Kingdom are by trapping and, more usually, by poisoning with strychnine. Both methods give cause for concern in that they involve a varying degree of cruelty. The use of strychnine and other poisons is also problematical in that moles dying on the surface may be the cause of secondary deaths in carnivores and birds of prey. In addition, in Britain an unknown but probably significant quantity of strychnine is used illegally to kill a variety of other species which are regarded, by some, as vermin.

Clearly, there is a need for alternative means of control. In England, MAFF have put a great deal of effort into developing poisons which work in a less-cruel manner and which are relatively specific to moles. Unfortunately, their efforts have so far met with only limited success.

As part of an ongoing study of mole behaviour we have looked at the possibility of repelling moles, by means of odorous chemicals, from areas where they are causing damage. To date we have carried out experiments with two types of substances, social odours produced by moles themselves and social odours produced by mustelid carnivores. The rationales behind the use of these two classes of compounds are quite different and thus the two approaches are best dealt with separately.

RESPONSES TO MUSTELID SOCIAL ODOURS

Introduction

Carnivores frequently mark their ranges with odorous faeces, urine and the secretions of skin glands. These substances also often impart a powerful odour to the animals themselves. It is usually assumed that such scent marking behaviour is adaptive in that the information conveyed by a scent mark increases the inclusive fitness of the animal making it. However, there is a price to be paid for gaining such advantage in that the predator's rank odour may give an early warning to prey of its presence, allowing them to take appropriate evasive action. For example, microtine voles actively avoid the scent of stoats *Mustela erminea*, weasels *Mustela nivalis* and red foxes *Vulpes vulpes* (Gorman, 1984; Stoddart, 1976; Dickman and Doncaster, 1984). Avoidance responses of this kind have been utilized in various attempts to control the damage done by mammalian herbivores. Thus 3-propyl-1,2-dithiolane and 2,2-dimethylthietane from stoat and mink, *Mustela vison*, anal sac secretions respectively have been shown to inhibit the feeding of snowshoe hares *Lepus americanus* (Sullivan and Crump, 1984) while wolverine, *Gulo gulo*, lynx, *Lynx lynx* and bobcat, *Lynx rufus*, faeces reduce feeding damage by snowshoe hares and black-tailed deer, *Odocoileus hemionus* (Sullivan, Nordstrom and Sullivan, 1985; Sullivan, 1986).

We have been encouraged by results such as these to investigate the responses of moles to the anal sac secretion of the weasel and to synthetic 2,2-dimethylthietane, a major product of the anal sacs of mink. Neither of these two mustelids are major predators of moles but the reactions of prey often seem to be to predator odours in general rather than to those of specific species with which they are familiar. Thus, for example, Orkney voles, *Microtus arvalis*, are repelled by the scent of stoats despite the fact that they have been isolated from these predators for at least 5000 years (Gorman, 1984) and it has been claimed that British field voles, *Microtus agrestis*, avoid tiger, *Panthera tigris*, urine, and red deer, *Cervus elaphus*, lion, *Panthera leo*, faeces (Stoddart, 1980).

Materials and methods

Our field studies, carried out in north-east Scotland, were designed to determine whether free-living moles actively avoid predator odours. The initial approach, based on Stoddart's (1976) work was to compare the catching efficiency of clean Frisian live-traps with that of traps tainted with weasel odour, with 2,2-dimethylthietane or with oil of peppermint. The peppermint treatment was included to control for the fact that moles may respond to an odour because it is novel, or noxious rather than because it signals predator.

The trapping was carried out over the period November to January inclusive with an equal trapping effort being expended for each of the four trap types. Only one trap was set in any one mole territory, and each was left in place for up to four days with daily resetting if a mole blocked off the trap with soil. Only one (4 day) attempt was made to catch any particular mole. At any one time equal numbers of each trap type were set in order to allow for any temporal changes in trapability. The traps were used for one treatment only in order to avoid cross-contamination of odours.

We have also carried out a single field trial designed to determine whether or not moles will avoid tunnels treated with predator odours. For our experiment we chose a mole which, from the dispersion pattern of its mole hills, appeared to have a home range consisting of two main areas connected by a narrow strip of raised, dry land passing between wet, boggy depressions. We caught the animal, an adult male, in December and attached to its tail a 2.1 g radio transmitting in the 173 MHz range (for details of attachment see Stone and Gorman, 1985). With the radio it was possible to locate the mole's position to an accuracy of ± 0.25 m.

The movements of the animal were traced by locating it every 10–15 min whilst it was out of its nest. Initially we plotted its movements over a period of six days. On each of these six days we followed the animal throughout one of its three activity periods (see below), advancing one period of each consecutive day. On the seventh day, while the mole was asleep in its nest, we placed a total of 30 ether-washed cigarette filters into the tunnels connecting the two parts of the range and at various points throughout the upper part of the range. The filters were introduced into the tunnels through small holes made with a metal probe and subsequently sealed with soil. We then continued to plot the movements of the mole for a further three days, again covering a different activity period on each day.

At the end of this, control, period an additional 30 filters, this time soaked in 2,2-dimethylthietane, were placed in the tunnels at approximately the same positions as before. The mole was then tracked, again over one activity period per day and advancing one period between days, for a further eight days at which point the radio failed.

Results

The trapability of moles in clean and treated traps are shown in Table 8.1. Clearly, the traps treated with weasel odour or with 2,2-dimethylthietane were markedly less efficient at catching moles than were the clean traps, while adding oil of peppermint appeared to have no effect. We conclude that moles detect and avoid both natural weasel scent and synthetic 2,2-dimethylthietane.

Table 8.1 The trapping efficiency of Frisian traps

Number of captures in:			
Clean traps	Peppermint traps	Weasel traps	Dimethylthietane traps
12	14	1	2

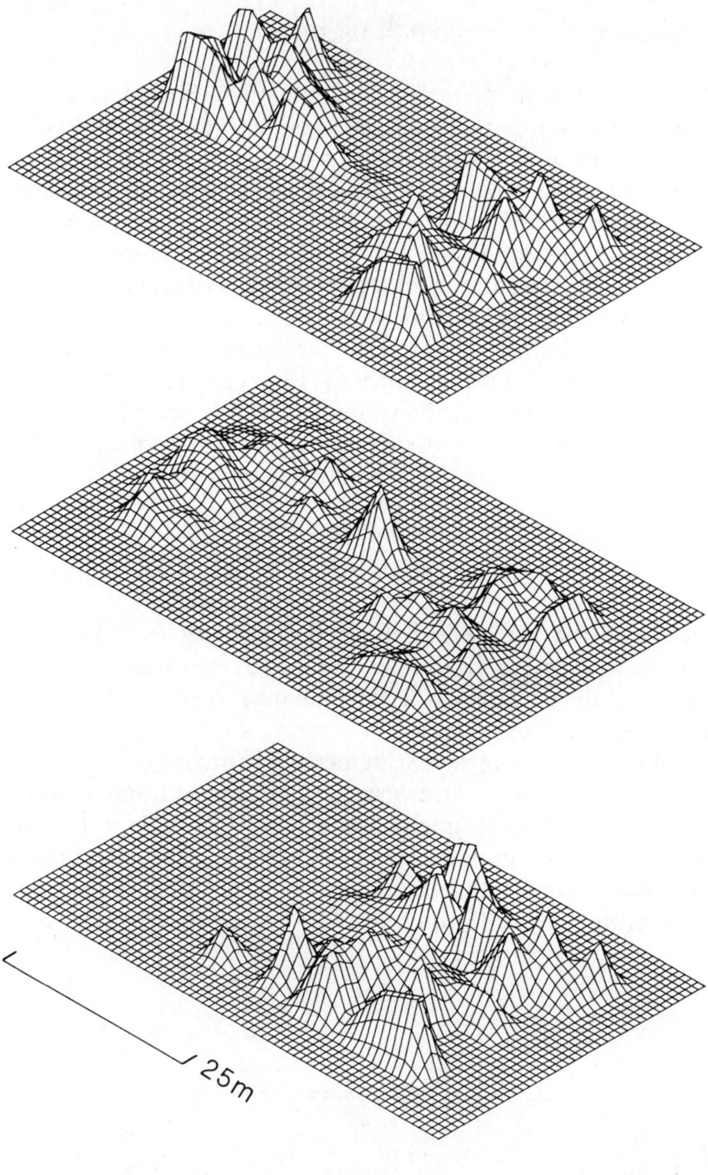

Figure 8.1 Three-dimensional representations of the use made by a mole of different parts of its home range. The plots were made by dividing the map into a matrix of 25 m^2 squares and counting the number of times the mole was located in each. The maps were drawn by SURFACE II (Sampson, 1978) as if viewed from the south-east with the observation point situated 10 000 map units from the centre of the matrix and 20° above the horizontal.

The upper map shows how the range was utilized prior to the experiment, the middle map shows the situation following the introduction of control stimuli, and the lower map the result of introducing dimethylthietane into tunnels. See text for further details.

Radio-tracking confirmed that the male mole did indeed have a range consisting of two main areas connected by, as far as we could detect, just three tunnels and with a single nest in the lower part of the range. The degree to which the mole used the different parts of its range during the three phases of the experiment are shown in Figure 8.1. In these 3-dimensional maps the vertical dimension reflects the number of times the animal was located in different parts of the range. During the first six days of tracking the mole used all parts (Figure 8.1, top), and continued to do so following the introduction of the clean, control cigarette filters (Figure 8.1, middle). However, once the filters treated with 2,2-dimethylthietane had been placed in the tunnels the mole ceased to forage in the upper area, and began to extend the network of tunnels in the lower, untreated part of the range (Figure 8.1, bottom).

RESPONSES TO MOLE SOCIAL ODOURS

To understand the rationale behind the next set of experiments it is important to appreciate that moles, despite living underground, are very much aware of the presence and movements of their neighbours and tend to avoid direct contact. We have a variety of evidence, based on radio-tracking data to support this statement. Throughout most of the year adult moles are sedentary and occupy a mosaic of home ranges (Figure 8.2; in this and in subsequent figures home ranges are depicted as minimum convex polygons because extensive analyses of a large number of ranges have shown that this model is as appropriate as any for describing mole ranges (Stone, 1986)). Generally, there is no overlap

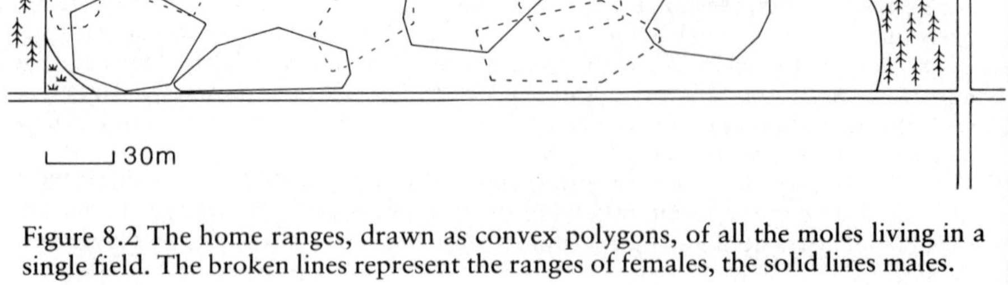

Figure 8.2 The home ranges, drawn as convex polygons, of all the moles living in a single field. The broken lines represent the ranges of females, the solid lines males.

between the ranges of neighbouring male moles, but each male's range overlaps with a varying number of female ranges and female ranges may also overlap those of other females (Figure 8.2). The overlap between the ranges of neighbours involves both the inter-digitation of separate tunnels in the soil column and the sharing of tunnels.

Why then do we say that individuals are aware of the presence and behaviour of other moles? The first line of supporting evidence is that if a mole is removed from its range neighbouring animals move, often rapidly, into the vacated area. As an example, consider Figure 8.3 which

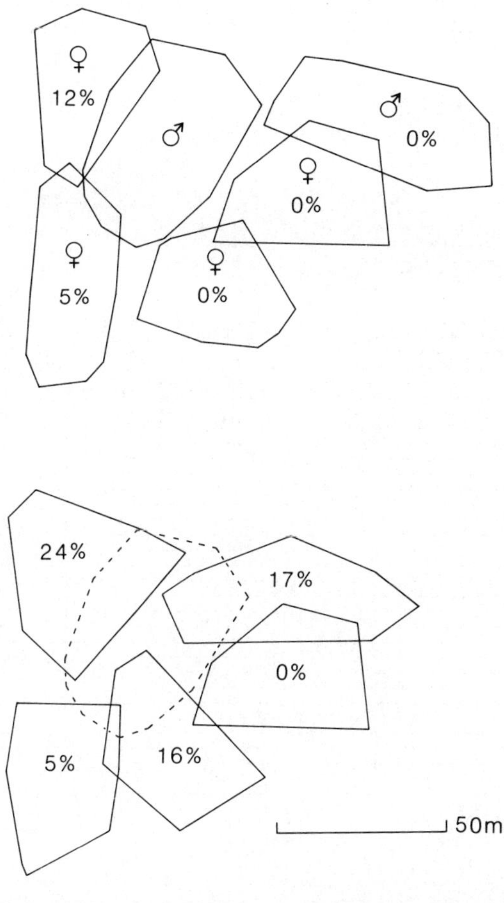

Figure 8.3 The home ranges, drawn as convex polygons, of six neighbouring moles. The upper and lower figures show the situation before and after the removal of the central male. The percentage values represent the proportion of the central mole's range that was overlapped by various neighbours, before and after its removal.

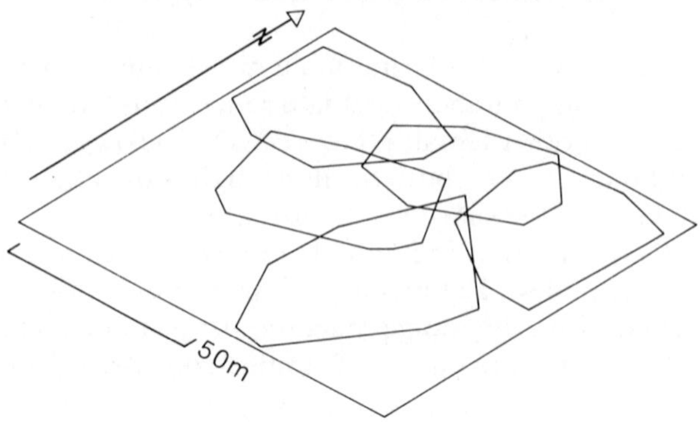

Figure 8.4 The home ranges, shown as convex polygons, of five neighbouring moles. The map is presented as if viewed from the south-east with the observer 20° above the horizontal.

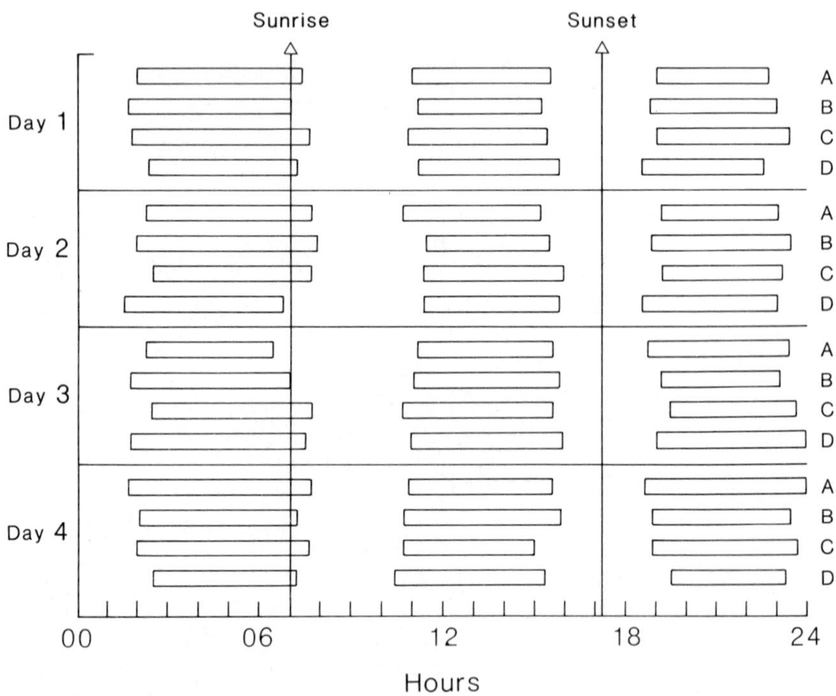

Figure 8.5 The activity patterns of four (A–D) of the moles in Figure 8.4 on four consecutive days. The closed bars represent the periods when the moles were asleep in their nests.

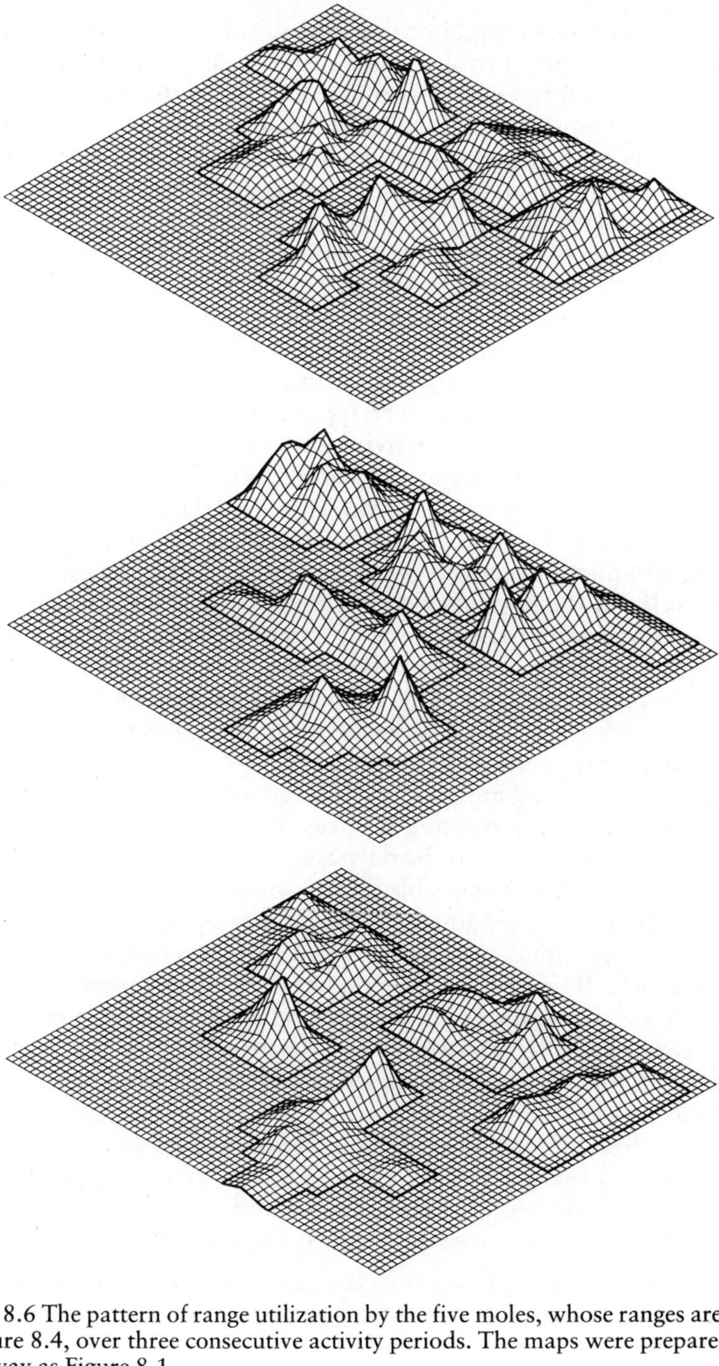

Figure 8.6 The pattern of range utilization by the five moles, whose ranges are shown in Figure 8.4, over three consecutive activity periods. The maps were prepared in the same way as Figure 8.1.

shows the ranges of six neighbouring moles. The percentage values in the figure indicate the proportion of the central male's range that was overlapped by the ranges of various neighbours. When the central male was trapped and removed, three neighbouring animals moved into and occupied parts of the empty range (Figure 8.3, bottom). The detection of the animal's demise was rapid, with the top left female, for example, starting to encroach within 15 hours.

The second line of evidence is that moles with overlapping ranges very rarely come into close contact with each other, thus avoiding the risks of overt aggression. Instead they display temporal avoidance with only one individual using the area held in common at any one time.

This is clearly shown by a group of five moles that we have tracked, each of which had a range overlapping with that of one or more neighbours (Figure 8.4). The activity patterns of four of these animals were monitored by continuous automatic recording of whether or not they were present in their nests. All four exhibited the typical mole activity pattern (Stone and Gorman, 1985) of three active foraging bouts per 24 hours, separated by periods of sleep spent in the nest (Figure 8.5). The activity cycles of the animals were markedly synchronized with all four leaving and returning to their nests at about the same time. During any one activity period the area of overlap between a particular pair of moles was used by one animal only. This is illustrated in Figure 8.6 which shows the way in which all five animals used their ranges during three consecutive activity periods.

Clearly neighbouring moles are well aware of each others' presence or absence and they avoid meeting in areas of range overlap. This implies that relevant information is being passed, by one means or another, between individuals. One possible route is via scent marking. Moles have paired preputial scent glands, containing both apocrine and sebaceous elements, which produce a secretion with an odour redolent of the general smell of a mole (Figure 8.7). This material becomes dispersed over the lower abdomen of moles and is also inevitably smeared onto the floor of the tunnel system as they move around. There is the potential, therefore, for signalling that an area is occupied and/or that a resident animal has passed this way in the recent past.

We have, therefore, conducted a number of field experiments designed to determine whether free-living moles do avoid preputial gland secretions. In addition, we have carried out a preliminary chemical analysis of the secretion together with statistical analyses to determine

Figure 8.7 The preputial glands of an adult male mole. (a) Gross dissection. (b) High power detail of the sebaceous component. (c) High power detail of the apocrine elements.

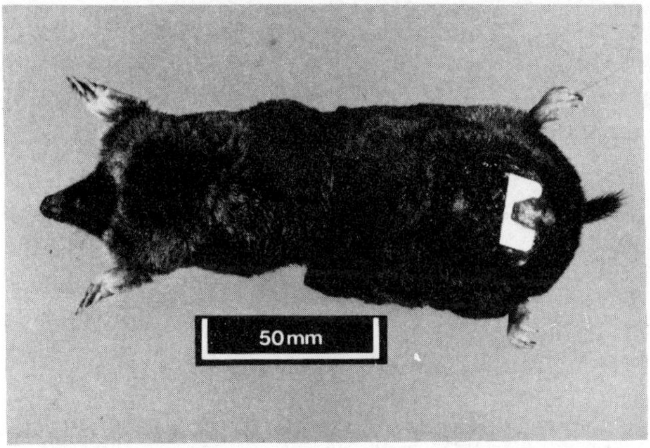

(a)

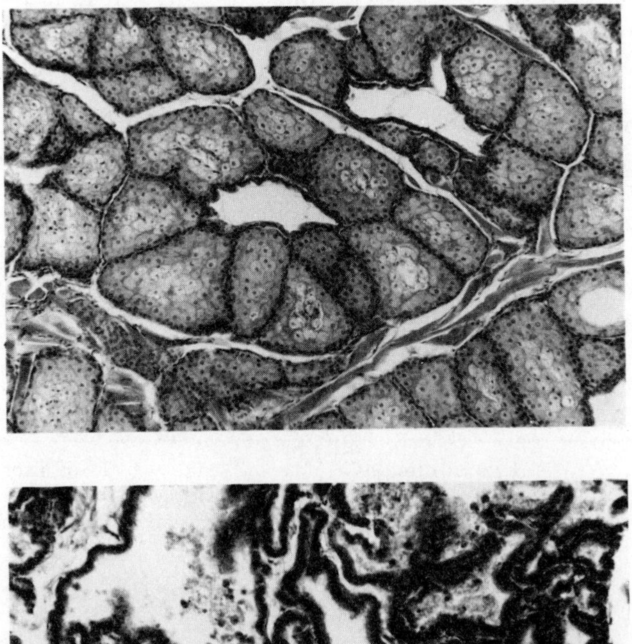

(b)

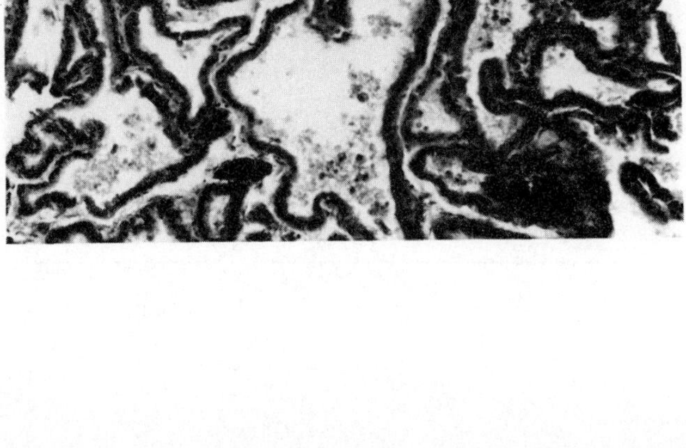

(c)

whether it contains information potentially useful in social behaviour; in particular whether there are individual differences and/or difference between the two sexes.

Materials and methods

Chemical analyses. The material was obtained by catching animals, throughout the year, in Duffus or half-barrel traps. The preputial glands were removed and placed in 100 μl of 'lipopure' hexane prior to analysis. Separations were carried out using a Pye-Unicam chromatograph equipped with dual-flame ionization detectors and a 5 ft × 0.25 in (1.5 m × 6.25 mm) (outside diameter) glass column packed with 3% dimethyl silicone gum on 100–120 mesh diatomite CQ. The analyses were temperature programmed from 170°C to 320°C at 1°C/min, using nitrogen at 60 ml/min as carrier gas. Peak areas were measured with an LDC 308 computer integrator.

Behavioural experiments. Our approach was to place cotton buds, either clean or smeared with preputial gland secretion, into the tunnels of radio-tagged moles and to record their responses when they encountered them. Five moles (three males and two females) were used in the trials. Each of them was tracked until we had an accurate knowledge of their tunnel systems and patterns of movements. It was then possible to place the stimuli into shallow tunnels, via a small hole, some 10–15 minutes before

Table 8.2 The response of free-living moles to clean, control cotton buds and to buds soaked in preputial gland secretion (see text for further details)

Mole number	Type of stimulus	Response: positive	negative	unclear
Male 1	strange male	15	2	3
	control	2	16	2
Male 2	strange female	15	1	1
	control	0	15	1
Male 3	familiar male	10	0	0
	control	1	10	0
Female 1	strange female	11	2	3
	control	0	16	0
Female 2	strange male	10	1	4
	control	1	7	2
	Totals			
	preputial secretions	61	6	11
	control	4	64	5
	Chi-square =	100.24	$p < 0.001$	

the mole was likely to arrive at that point. The reaction of the mole was scored as one of three categories: (1) a clear retreat back in the direction from which it had come; (2) a negative response if it moved past the stimulus along the tunnel; (3) an unclear response where we were not completely sure as to what had happened. The secretions used in the trials came from animals which had been trapped. Their preputial glands were removed immediately after death and stored at −25°C. All of the donor animals were strangers to the radio-tagged moles with one exception, a male who had been an immediate neighbour of male number 3.

A total of 75 control and 78 experimental presentations were made (Table 8.2). Only one presentation was made to a particular mole on any one day and the stimulus was then removed and the tunnel repaired. Subsequent presentations were made in different parts of the range.

Results

A typical chromatogram is shown in Figure 8.8. Differences in odour between different social classes are unlikely to be based on the possession of unique chemicals but rather upon differences in the relative concentrations of the components of a complex chemical mixture (see e.g. Gorman, 1976). The preputrial secretion is such a mixture with the same set of peaks, as identified by similar retention times, being present in all animals.

In order to detect possible differences between the secretions of individuals or between the two sexes 60 peaks were chosen for statistical analysis. These peaks, which were clearly resolved on all chromatograms are numbered 1–60 on Figure 8.8. It must be stressed that these particular peaks were chosen to be representative of the whole odour profile; there is no way of knowing at this stage whether they are more or less biologically important than the other peaks present. Our initial statistical approach has been to express, for each individual animal, the area under each peak as a percentage of the total area under all 60 peaks. This value we refer to as the peak proportion.

The mean peak proportions for 60 adult males and 58 adult females are shown in Figure 8.9. The coefficients of variation were large for all peaks indicating substantial individual differences in chemical profiles.

In order to detect any differences between the sexes a one-way analysis of variance was carried out for each peak in turn. The analyses revealed that there were 10 peaks which varied significantly more between, than within the sexes (Figure 8.9). There is, therefore, the potential for moles to be able to discriminate between the secretions of different individuals and to recognize the sex of the animal. However, whether or not they can do so is, as yet, unknown.

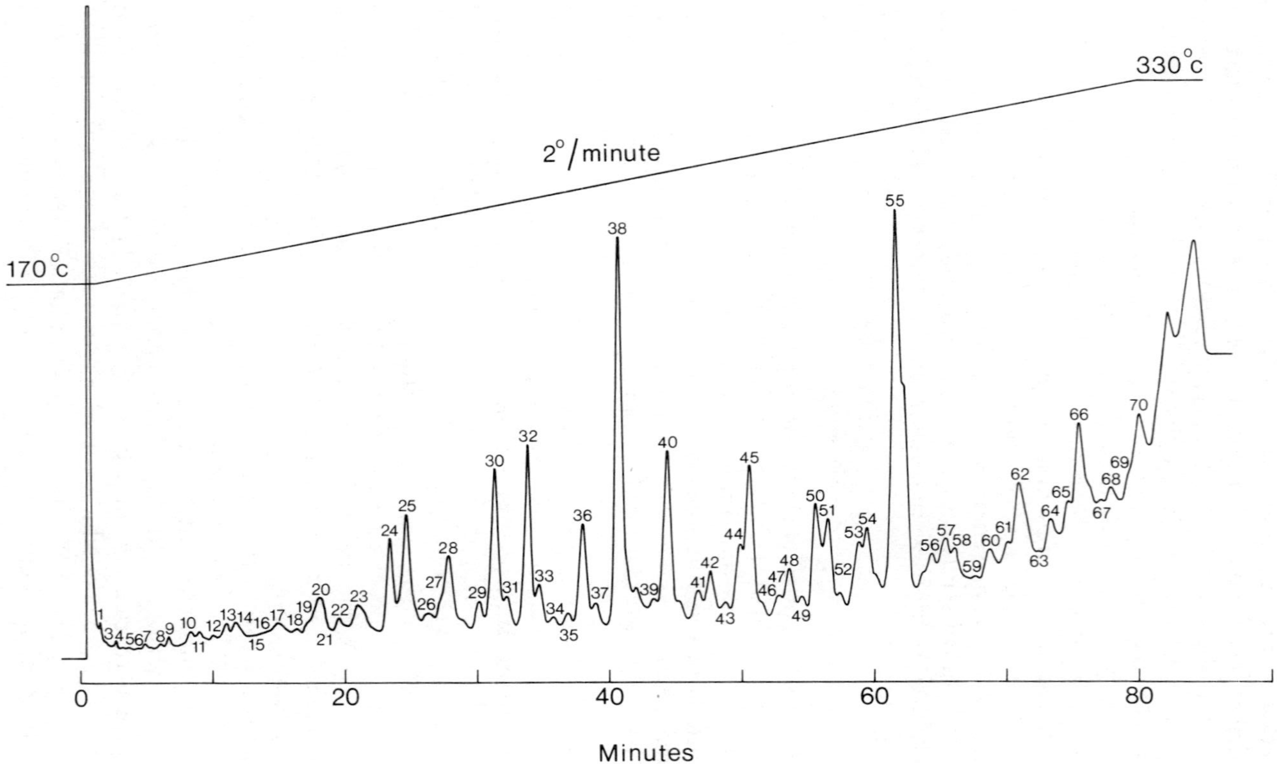

Figure 8.8 A typical gas-liquid chromatogram of the preputial secretion of a mole. The temperature programme used in the analysis is shown on the figure. The numbers indicate the peaks used in the statistical analysis of individual and sexual differences.

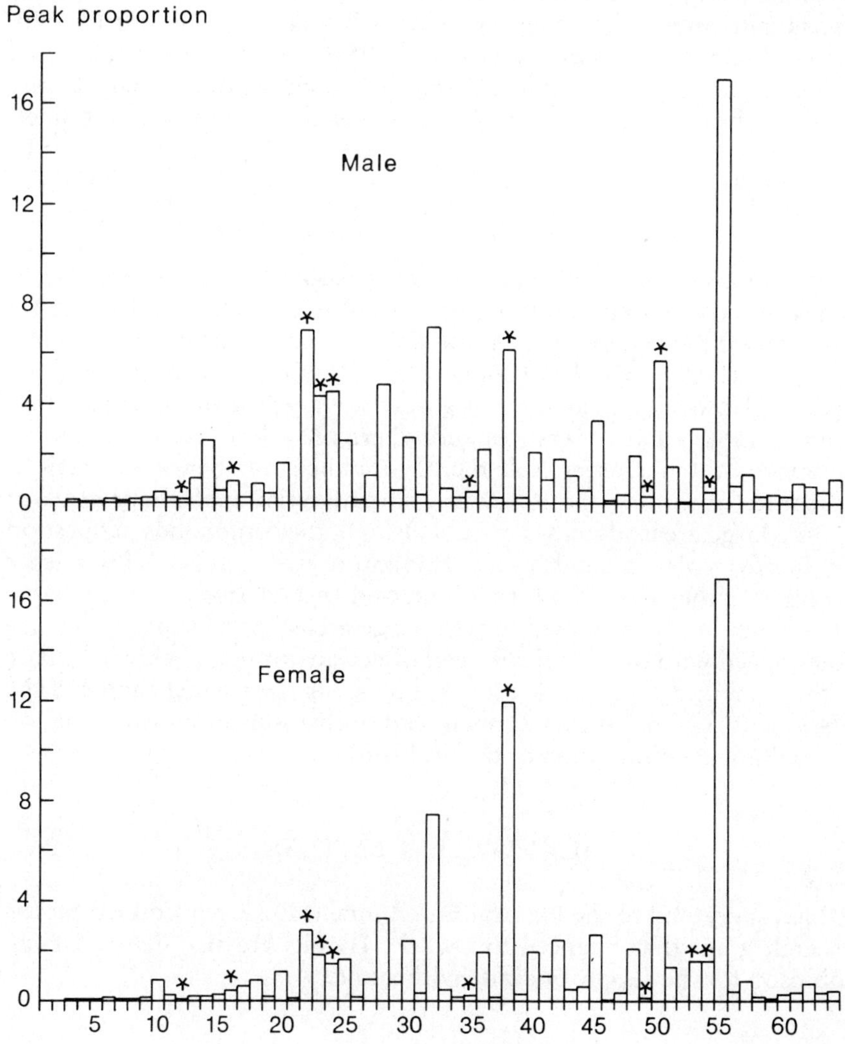

Figure 8.9 Mean peak proportions for the preputial secretion of 60 adult male and 58 female moles. The asterisks indicate peaks that varied significantly more between than within the sexes.

The reactions of the free-living moles to preputial gland secretions are shown in Table 8.2. The responses were very marked with moles almost always retreating when they encountered a cotton bud treated with preputial secretion, regardless of the sex of the donor or whether it had been a neighbour or an unknown animal. In contrast the animals usually passed by the control stimuli and continued on their way along the tunnel.

DISCUSSION

We have been able to show that moles are to some degree repelled both by conspecific odours and by those produced by mustelid carnivores. Are these compounds likely to be of practical use in mole control? In general we fear not! They may have some role to play in deterring moles from small, high amenity areas, such as gardens, where the owner is reluctant to apply more robust measures and where the cost of his/her labour is unimportant. On a larger scale there would be a number of very serious difficulties, including the following. The material and labour costs of treating large areas of tunnels would be high, the compounds in question are highly volatile, and even if incorporated into a slow-release formulation they would have to be renewed on a relatively frequent basis. If the compounds were used to treat large areas it is very likely that the moles, faced with the chemicals in all directions would quickly habituate to them or that they would simply block off the treated tunnels. The subsequent digging of new tunnels may well result in an exacerbation rather than in a diminution of the problem!

ACKNOWLEDGEMENTS

M.L.G. is grateful to the Natural Environment Research Council for the research grant that financed this work. Thanks are also due to Dr Ian Robinson for the supply of dimethylthietane.

REFERENCES

Dickman, C.R. and Doncaster, C.P. (1984) Responses of small mammals to red fox (*Vulpes vulpes*) odour. *J. Zool. Lond.*, **204**, 521–31.

Gorman, M.L. (1976) A mechanism for individual recognition by odour in *Herpestes auropunctatus* (Carnivora: Viverridae). *Anim. Behav.*, **24**, 141–5.

Gorman, M.L. (1984) The response of prey to stoat (*Mustela erminea*) scent. *J. Zool. Lond.*, **202**, 419–23.

Sampson, R.J. (1978) *Surface II graphics system.* Kansas Geological Survey, Lawrence, Kansas.

Stoddart, D.M. (1976) Effect of the odour of weasels (*Mustela nivalis* L.) on trapped samples of their prey. *Oecologia,* **22**, 439–41.

Stoddart, D.M. (1980) *The Ecology of Vertebrate Olfaction.* Chapman and Hall, London and New York.

Stone, R.D. (1986) The social ecology of the European mole *Talpa europaea* L. and the Pyrenean desman *Galemys pyrenaicus* G.: a comparative study. Unpublished PhD thesis. University of Aberdeen, Scotland.

Stone, R.D. and Gorman, M.L. (1985) Social organisation of the European mole (*Talpa europaea*) and the Pyrenean desman (*Galemys pyrenaicus*). Mammal Rev., **1**, 35–42.

Sullivan, T.P. (1986) Influence of Wolverine (*Gulo gulo*) odor on the feeding behavior of snowshoe hares *Lepus americanus. J. Mamm.*, **67**, 385–8.

Sullivan, T.P. and Crump, D.R. (1984) Influence of mustelid scent gland compounds on the suppression of feeding by snowshoe hares *Lepus americanus. J. Chem. Ecol.*, **10**, 1809–21.

Sullivan, T.P., Nordstrom, L.O. and Sullivan, D.S. (1985) Use of predator odours as repellants to reduce feeding damage by herbivores. 1. snowshoe hares *Lepus americanus. Chem. Ecol.,* **11**, 905–19.

Nine

The control of red and sika deer populations in commercial forests

P.R. RATCLIFFE

SUMMARY

Red and sika deer populations are currently increasing in density and expanding their range in many Scottish forests. These populations are responsible for damage to commercial tree crops, and, in some areas, are apparently modifying vegetation structure and species diversity. Interbreeding between red and sika deer is occurring in some areas, especially where range extension of either species is occurring. All of these problems require improved population control.

The economic significance of damage caused to commercial tree crops is difficult to quantify, owing to the uncertainty surrounding the potential for the tree to recover following injury. But, in some circumstances browsing can reduce early height increment thereby increasing establishment costs. Bark stripping injuries can result in down grading of timber from high quality structural grades to pulp or fire wood.

The estimated population of red deer in Scottish forests is being under-exploited by about 3500 deer each year and this also can be interpreted as an economic loss. The reasons for the inability to kill sufficient deer are associated with the dense nature of forest habitats and the constraints imposed by winter weather, close-seasons and legal methods of control. It seems likely that deer numbers will continue to rise, causing increasing economic loss and environmental damage, until density-dependent factors limit further increase.

ORIGINS AND THE STATUS OF POPULATIONS

Red deer

Red deer (*Cervus elaphus*) are indigenous to Great Britain and occur in the fossil record from the mid-Pleistocene onwards when they were associated with mixed coniferous/deciduous forests, oak forests and relatively open woodland (Lister, 1984). They may have been present throughout the last interglacial period but only occurred sporadically during the last (Devensian) glaciation when vegetation cover fluctuated widely.

During the postglacial expansion of birch/pine forest, red deer became the dominant large herbivore and they were well distributed throughout Great Britain. This persisted throughout the transition to predominantly deciduous forest at about 7000 BP. The present-day distribution is due largely to man's removal of forest cover since *c.* 5000 BP (Lister, 1984).

Deforestation and persecution probably resulted in the extinction of all native red deer in the lowlands by the seventeenth century (Ritchie, 1920) and in the Scottish highlands numbers were probably at their lowest at the end of the eighteenth century (Staines and Ratcliffe, 1987). Afforestation during the last thirty years has reversed the trend but some 5 million ha (75%) of Scotland remains devoid of trees (Ratcliffe, 1984a).

During the nineteenth century an increasing interest in red deer stalking led to deliberate attempts to increase numbers in Scotland. Females were protected and numbers rose more or less continuously to the present estimated level of 350 000 (Staines and Ratcliffe, 1987). Commercial forests established during the last 30 years or so have undoubtedly contributed to the increase in range and numbers, and about 50 000 red deer are estimated to occupy commercial forests in Scotland (Ratcliffe, 1987a).

Sika deer

Sika-like deer were present in the middle Pleistocene in Eastern Asia and have been represented there until the present time (Lister, 1984). The current natural range of sika deer (*Cervus nippon*) consists of deciduous and mixed woodlands and in Europe, like the red deer, its distribution is apparently limited by the period of snow-lie and cold windy conditions (Whitehead, 1972).

Sika deer were introduced to Great Britain from 1860, but the precise origins of the introduced deer are unknown. Their distribution and current status in Great Britain were reviewed by Ratcliffe (1987b). There has been a considerable increase in their range recently, especially in Scotland where some of the isolated introductions of the early part of this century are now coalescing to form contiguous populations (Ratcliffe, 1987b). This increase, even more so than red deer, has been a direct consequence of the spread of commercial afforestation.

Hybrids

Hybridization between red and sika deer appears to be increasing, and in Wicklow, Ireland, and in the south Lake District in England complete introgression has occurred resulting in populations composed entirely of hybrids (Ratcliffe, 1987b). Continuing hybridization is considered to present a threat to the genetic integrity of native red deer.

THE NEED FOR CONTROL

Damage caused by red, sika, and red × sika hybrids to commercial tree crops is apparently similar. Young trees can be heavily browsed resulting in a considerable delay in the trees reaching maturity. At a later stage bark-stripping can cause timber degrade, fungal entry, snapping of the main stem or death of the tree (Mitchell, Staines and Welch, 1977). Marauding onto agricultural land by red deer may be influenced by high densities in adjacent woodlands.

Damage by deer to commercial trees is widespread and occurs throughout the range of each species of deer (Tee, Rowe and Pepper, 1985). Vulnerability to damage depends upon the tree species (Staines and Welch, Chapter 11) and on their age or growth stage (Figure 9.1).

The economic significance of damage is difficult to quantify due to the uncertainty surrounding the potential for trees and other plants to recover following injury (Burdekin and Ratcliffe, 1985), and to the difficulty in evaluating the non-cash benefits of plant conservation. The importance of site quality in relation to the ability of plants to recover from damage and effect compensatory growth requires further investigation (Burdekin and Ratcliffe, 1985).

Comparisons between undamaged Sitka spruce, *Picea sitchensis* trees and those which had their leading shoot removed by deer, suggest that

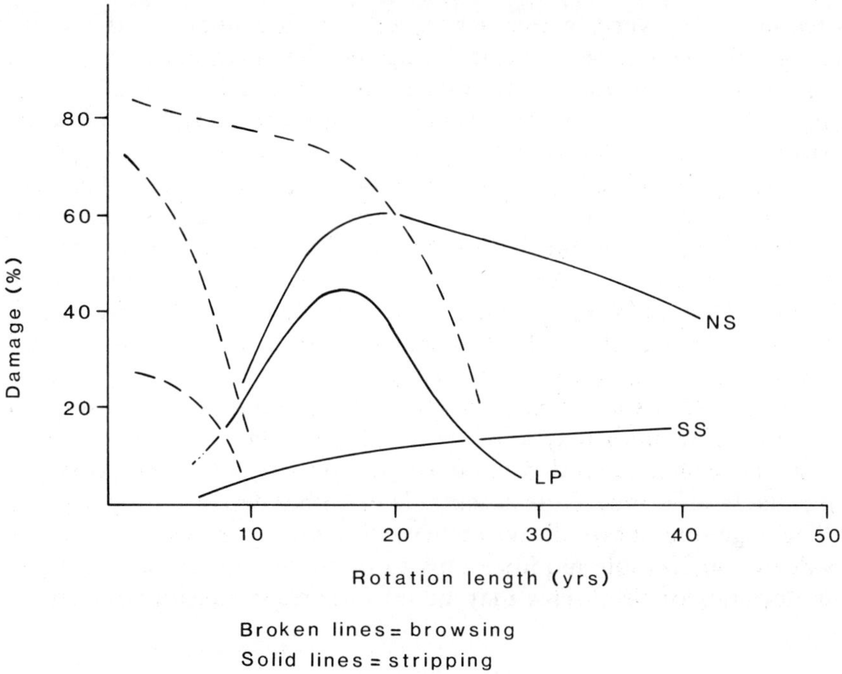

Figure 9.1 Vulnerability to damage (from Burdekin and Ratcliffe, 1985). NS – Norway spruce; SS – Sitka spruce; LP – Lodge pole pine.

trees on relatively fertile sites in western Scotland rapidly compensated for damage by increased growth. Trees which had their leading shoot removed by deer were not significantly smaller than their undamaged neighbours after 3 years (Tee, L.A. unpublished). However, on less-fertile sites in Northumberland, Sitka spruce trees did not recover from leader damage by roe deer and fencing has become accepted practice to protect commercial tree crops in this area. Recent work on Norway spruce, *Picea abies*, has shown that early heavy browsing on lateral branches, leaving the leading shoot intact, significantly reduced height growth over the following eight years (Pepper, H.W. and Hewison, M. unpublished). Kampmann (1983) obtained similar results for roe deer in Germany.

Browsing of leading shoots can also lead to the production of multi-leaders (Staines and Welch, 1984). When the final crop is intended for pulpwood, or if the number of affected trees is low, this may be unimportant, but it will reduce the value of trees destined for structural timber.

It appears that forest managers may over-react to early browsing damage by deer, especially where site quality is high and trees have a high potential for recovery. In areas where recovery does not occur the costs of damage can be assessed by calculating the change in net present value associated with the additional time required to reach economic rotation length. However, there may be other losses due, for example, to the added costs of weed control during the early stages of growth.

Bark-stripping damage occurs later in the rotation and when large areas of bark are removed this is likely to be of greater economic significance than browsing. Sitka spruce is less likely to be damaged in this way than other commercially grown conifers (Staines and Welch, Chapter 11) while Norway spruce is particularly susceptible (Figure 9.1). The value of losses due to bark-stripping injuries is difficult to assess or predict, as trees can recover. Little is known of the economic significance of the removal of small patches of bark. However, the callusing and occlusion of wounds may allow pathogens to enter and almost always result in some structural degrade of the timber. A low proportion of damaged trees is apparently acceptable in timber destined for pulping.

The browsing of broadleaved tree seedlings and herbaceous plants can result in considerable modifications to plant communities at relatively low densities of deer. This may be an important consideration in the

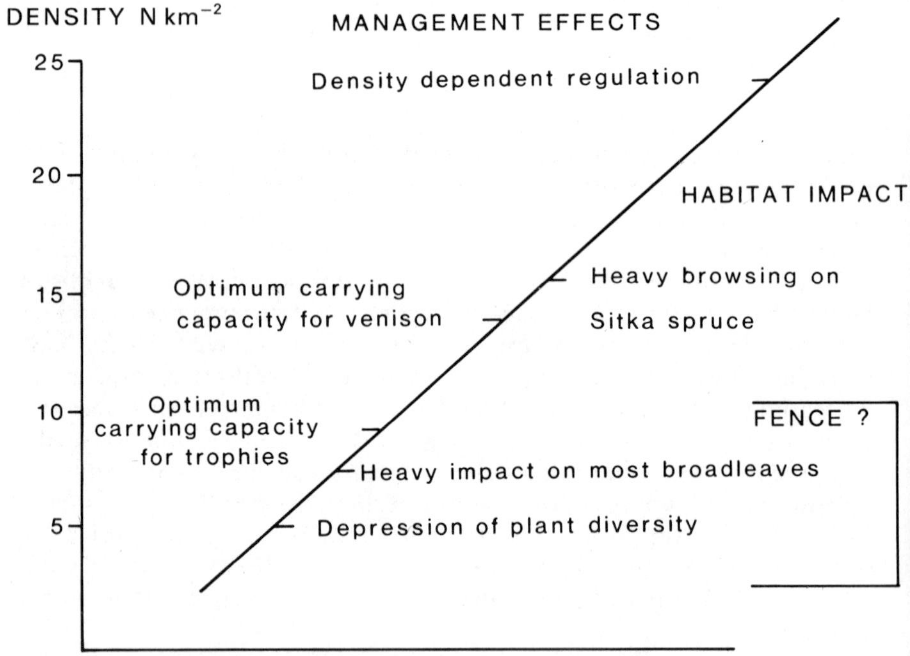

Figure 9.2 Carrying capacity for deer (from Ratcliffe, 1988).

management of vegetation and their dependent animals for conservation reasons (Ratcliffe, 1988).

Deer management for maximum sustained yield of venison or trophy antlers demands progressively lower densities of deer and quality can only be achieved with a reduction in numbers (Ratcliffe, 1988). It is unfortunate that the levels of density at which density dependent factors begin to regulate population growth and limit population size are well above the density at which crop damage occurs (Figure 9.2).

CONTROL

Control means to regulate. It is possible to control damage levels and at the same time have little effect on population density or numbers. In this sense population control means the ability to regulate animal numbers or densities. Control of deer can only be achieved legally in Great Britain by killing deer with certain prescribed firearms within clearly defined seasons. Control can only be demonstrated to have occurred when confident estimates of population size before and after culling operations indicate a reduction. Recent work suggests that the population dynamics and densities of sika deer are similar to those of red deer, but further research is needed (Ratcliffe, 1987b and unpublished data).

In Galloway, south west Scotland, direct counts of red deer on a largely unafforested area indicated a population of about 1500 deer in 1965. Management objectives were set to reduce the population to 500 deer to reduce the impact on the forest and improve the quality of the deer. Direct censuses were continued and by 1976 these indicated that the objective had been achieved. However, during this period a substantial amount of the range had been afforested and dense thickets were present over large areas, which dramatically reduced the efficiency of direct census. More importantly, the numbers of deer being killed were much higher than a total population of 500 could sustain (Ratcliffe, 1987c). Clearly, the estimate of 500 deer was inaccurate.

Ratcliffe (1987a) projected the changes in this population from 1965 based upon the 1965 counts, and a knowledge of recruitment rates and annual culls. The projection was checked independently using vantage-point counts (Ratcliffe, 1984b; 1987c) to estimate densities.

Projections indicated a rise from about 1500 in 1965 to about 2000 in 1976 and numbers have continued to increase to about 4000 deer in 1986. Little is known of the relationship between density and rate of dispersion in red deer and it is not clear whether this number of deer should be related to the original area occupied by the deer in 1965–76, or whether they have dispersed to occupy a wider area. However, independent estimates suggest identical densities to the 9.0 deer km^{-2} derived from the projected figure of 4000 deer. This density is rather low compared with other populations which again suggests low rates of dispersion. Additionally, subjective observations and cull data from

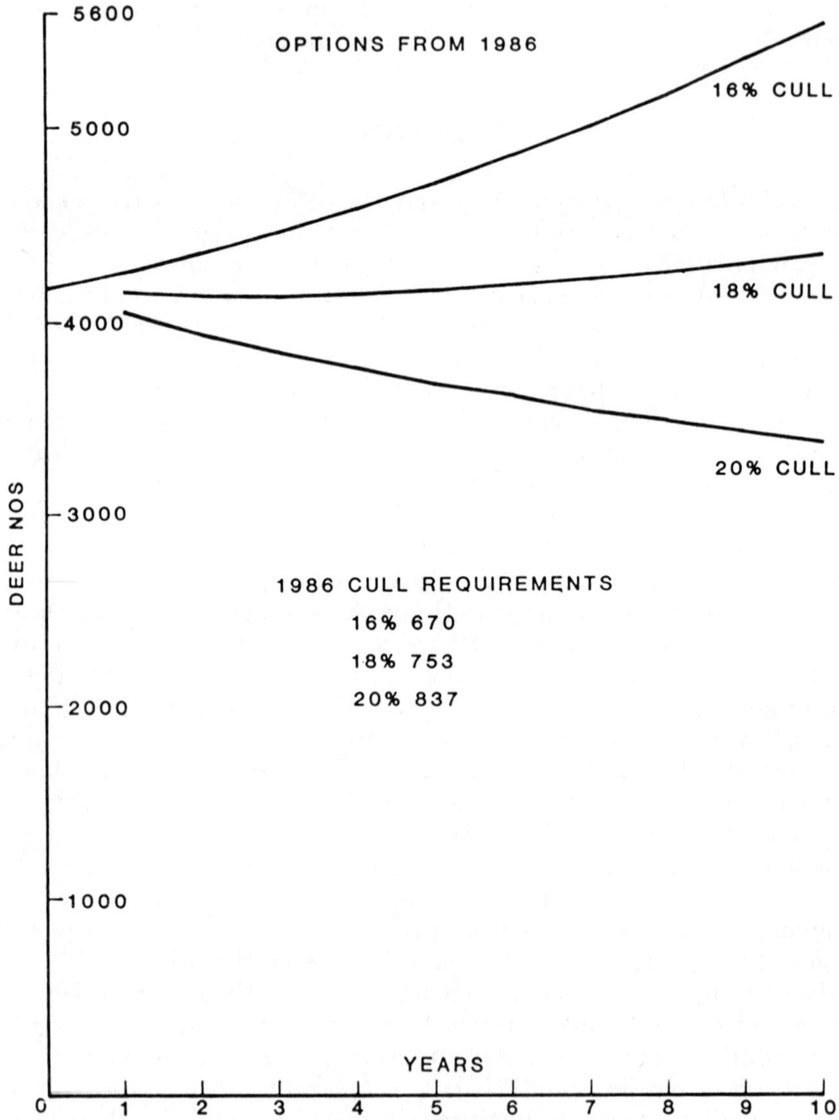

Figure 9.3 Red deer population changes – Galloway (from Ratcliffe, 1987a).

surrounding areas do not suggest large or increasing dispersion from the central area.

Underestimates of red deer numbers have occurred elsewhere, with actual densities of 4 times previous estimates being common, with one area underestimating numbers by 17 times (Ratcliffe, 1987a). The reproductive performance of red deer in Galloway is the highest so far recorded in Scotland with over 85% of yearling females becoming pregnant. (Ratcliffe, 1984b). Figure 9.3 illustrates the effects of three levels of cull on population growth in this population.

Clearly many populations are performing at levels below that of Galloway and these would not sustain culls as high as the 18% proposed for Galloway (Figure 9.3) (Ratcliffe, 1987a). A cull of 16% is recommended for many open-range populations with poor survivorship and growth rates (Red Deer Commission, 1981). Most forest populations are considered to fall between these levels and some exhibit reproductive rates similar to Galloway (Ratcliffe, 1984b).

Culling requirements have so far been discussed in terms of total numbers, but it is of great importance that sufficient breeding-age females are included in the cull. In Galloway, from the total cull of 837 recommended in 1986 (Figure 9.3), 350 must be breeding-age females, representing 40% of the cull, to prevent an increase in numbers.

Tables indicating the levels of cull needed to achieve stability (zero growth), based on the proportion of juveniles in the annual kill and the rates of reproduction and survivorship (Wegge, 1973), were modified to include densities (Ratcliffe, 1987c). Table 9.1 indicates the required cull of breeding-age females, males, and juveniles ($N\,km^{-2}$) to achieve stability.

The upper and lower estimates for the cull of breeding-age females to achieve zero population growth (Table 9.1) are compared with the actual culling achievement for a range of populations considered to reflect the range of densities and performance encountered in Scottish commercial forests (Figures 9.4–9.8) (Ratcliffe, 1987d).

Many populations are clearly increasing in numbers and can sustain much greater culls than are actually being imposed. In Galloway an annual cull of *c.* 11% is allowing the population to continue to increase (Figure 9.5). A cull of 18% is necessary to prevent this. Only in Cowal (Figure 9.4) is there evidence that culling is having some impact on population growth, though even this population is increasing slowly.

Assuming that an average sustained annual level of exploitation of 18% is possible, then the actual culls of a 11% being achieved represent underculling of *c.* 7%. Assuming that there are 50 000 red deer resident in Scottish forests, this amounts to 3500 deer each year (Ratcliffe, 1987c). Little is known of the relationships between the level of damage and increasing deer numbers, but the potential for high damage is obviously

Table 9.1 Cull of red deer required to achieve population stability (deer km^{-2})

Juv. % in kill	Adult	Spring density (deer/100 ha) 5 M	5 F	5 C	10 M	10 F	10 C	15 M	15 F	15 C	20 M	20 F	20 C	25 M	25 F	25 C
M:F	M:F							High performance, low mortality								
20:20	1:1	0.25	0.28	0.07	0.51	0.56	0.15	0.76	0.84	0.22	1.02	1.12	0.29	1.27	1.40	0.37
25:25	1:1	0.29	0.34	0.11	0.57	0.69	0.22	0.86	1.03	0.33	1.14	1.37	0.43	1.43	1.71	0.54
30:30	1:1	0.32	0.38	0.14	0.63	0.75	0.28	0.95	1.14	0.42	1.26	1.51	0.55	1.58	1.89	0.69
35:35	1:1	0.34	0.45	0.19	0.68	0.89	0.38	1.03	1.34	0.58	1.37	1.78	0.77	1.71	2.23	0.96
40:40	1:1	0.37	0.49	0.24	0.75	0.97	0.49	1.12	1.46	0.73	1.49	1.94	0.97	1.87	2.43	1.21
45:45	1:1	0.39	0.55	0.31	0.78	1.09	0.63	1.17	1.64	0.94	1.56	2.19	1.25	1.95	2.73	1.56
50:50	1:1	0.42	0.58	0.38	0.83	1.16	0.76	1.25	1.74	1.13	1.66	2.33	1.51	2.08	2.91	1.89
55:55	1:1	0.42	0.64	0.47	0.85	1.27	0.93	1.27	1.91	1.40	1.70	2.55	1.87	2.12	3.18	2.33
60:60	1:1	0.43	0.69	0.56	0.86	1.37	1.12	1.29	2.06	1.68	1.71	2.74	2.25	2.14	3.43	2.81
65:65	1:1	0.43	0.73	0.67	0.86	1.46	1.33	1.29	2.19	2.00	1.72	2.92	2.66	2.15	3.65	3.33
70:70	1:1	0.43	0.77	0.77	0.85	1.52	1.53	1.28	2.30	2.31	1.70	3.07	3.08	2.13	3.83	3.85
75:75	1:1	0.41	0.81	0.90	0.81	1.63	1.81	1.22	2.44	2.71	1.63	3.26	3.62	2.04	4.07	4.52
								Low performance, low mortality								
20:20	1:1	0.26	0.26	0.71	0.52	0.52	0.14	0.79	0.79	0.21	1.05	1.05	0.28	1.30	1.30	0.35
25:25	1:1	0.30	0.30	0.10	0.59	0.59	0.19	0.89	0.89	0.29	1.19	1.19	0.38	1.49	1.49	0.48
30:30	1:1	0.33	0.33	0.12	0.66	0.66	0.24	0.99	0.99	0.37	1.32	1.32	0.49	1.65	1.65	0.61
35:35	1:1	0.36	0.36	0.16	0.73	0.73	0.31	1.09	1.09	0.47	1.46	1.46	0.63	1.82	1.82	0.78
40:40	1:1	0.40	0.40	0.20	0.80	0.80	0.40	1.20	1.20	0.60	1.60	1.60	0.80	2.00	2.00	1.00
50:50	1:1	0.46	0.46	0.30	0.93	0.93	0.60	1.39	1.39	0.90	1.85	1.85	1.20	2.30	2.30	1.50
								Low performance, high mortality								
20:20	1:1	0.32	0.32	0.11	0.63	0.63	0.22	0.95	0.95	0.33	1.27	1.27	0.44	1.58	1.58	0.55
25:25	1:1	0.35	0.35	0.15	0.71	0.71	0.29	1.06	1.06	0.43	1.41	1.41	0.58	1.76	1.76	0.72
30:30	1:1	0.39	0.39	0.18	0.77	0.77	0.36	1.16	1.16	0.54	1.54	1.54	0.72	1.93	1.93	0.90
35:35	1:1	0.43	0.43	0.24	0.85	0.85	0.48	1.28	1.28	0.72	1.71	1.71	0.96	2.14	2.14	1.20
40:40	1:1	0.46	0.46	0.30	0.93	0.93	0.60	1.39	1.39	0.90	1.85	1.85	1.20	2.31	2.31	1.50

M = male; F = female; C = calf; Juv. = all 1 and 2 year olds. After Ratcliffe (1987c).

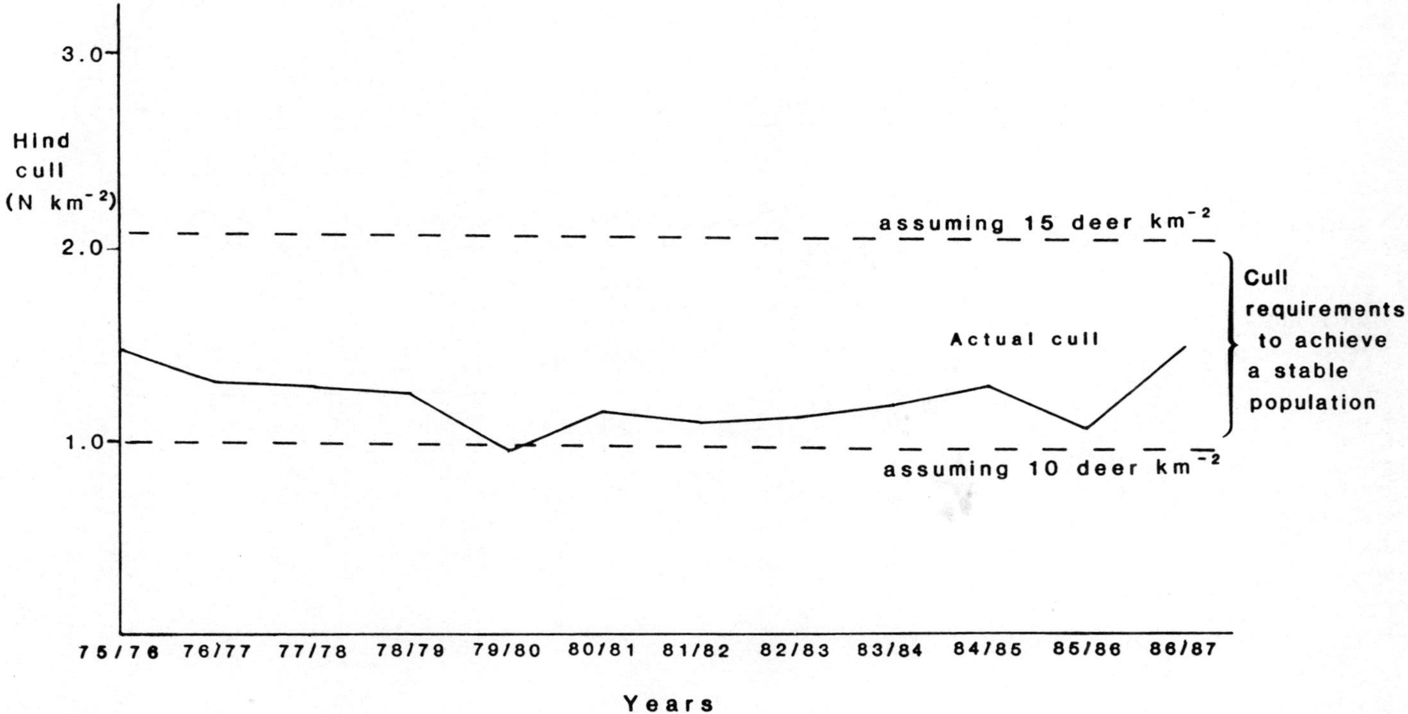

Figure 9.4 Red deer hind culling – Cowal. This assumes high reproductive performance and low mortality.

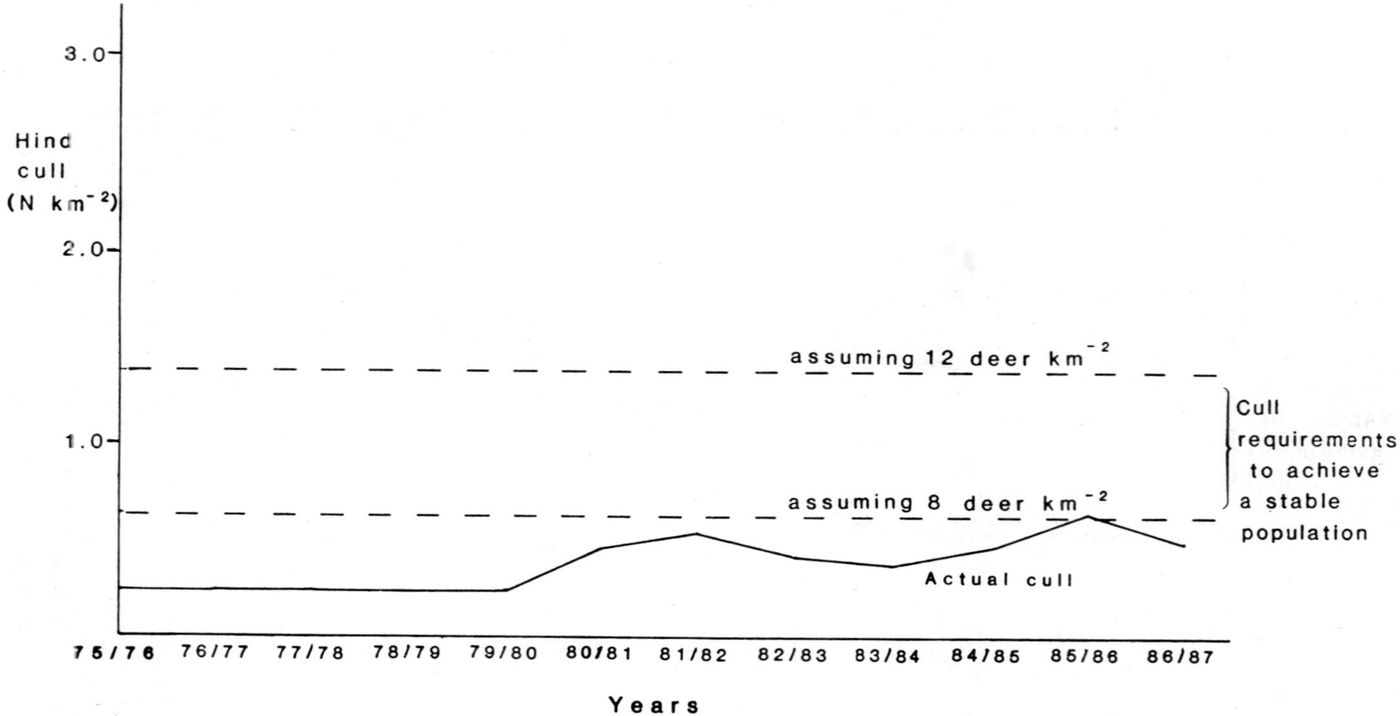

Figure 9.5 Red deer hind culling – Galloway. This assumes high reproductive performance and low mortality.

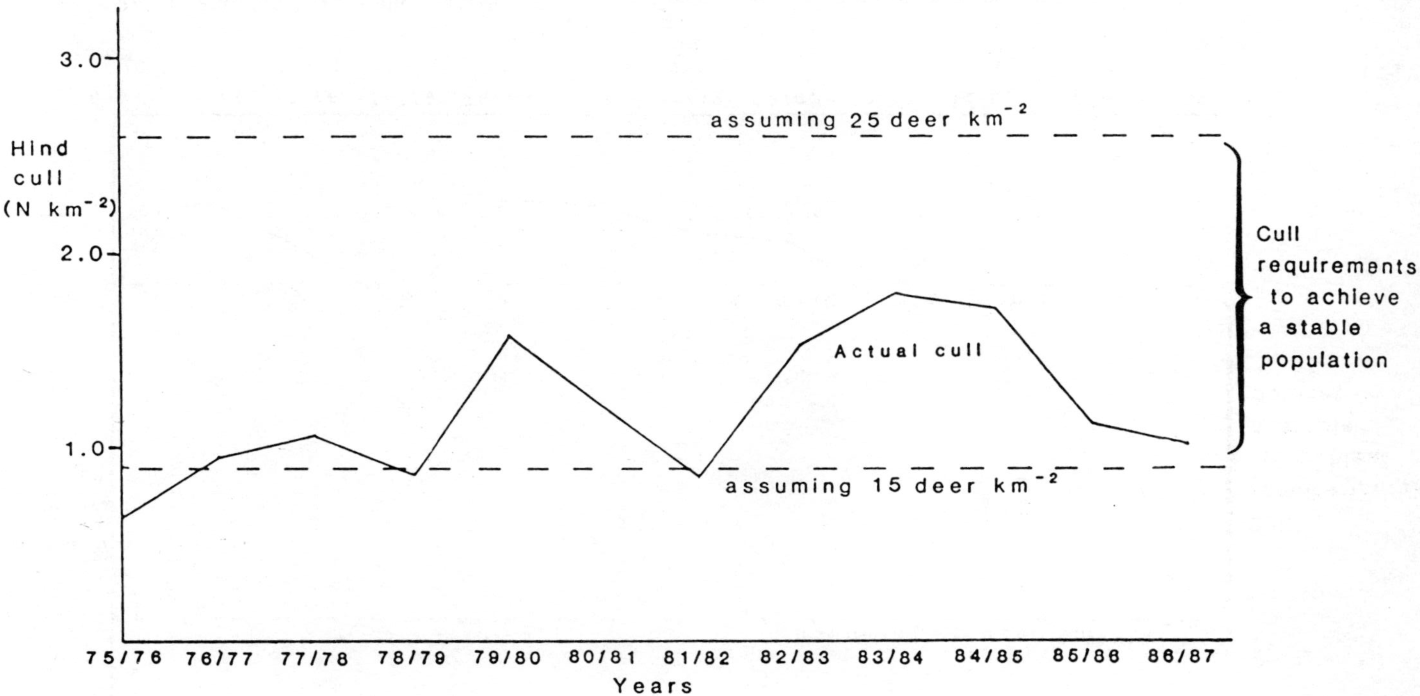

Figure 9.6 Red deer hind culling – Strathyre. This assumes low reproductive performance and low mortality.

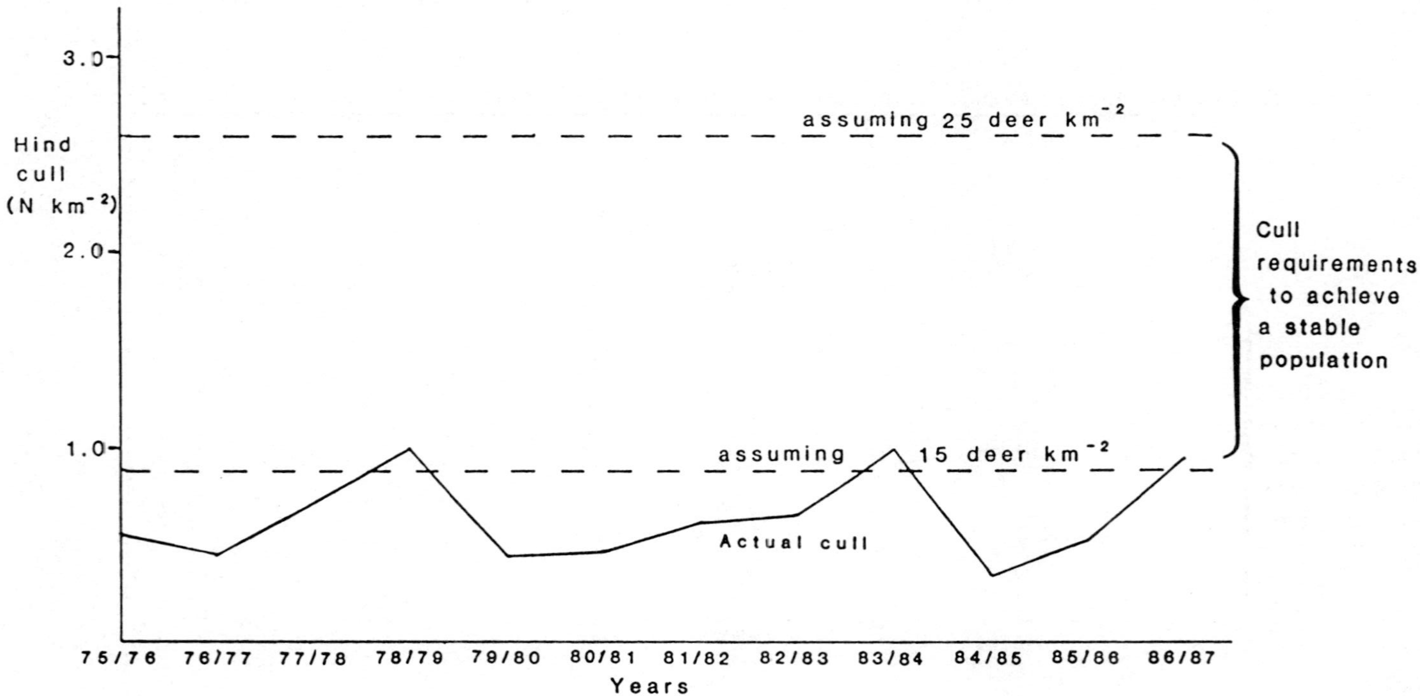

Figure 9.7 Red deer hind culling – Achnashellach. This assumes low reproductive performance and low mortality.

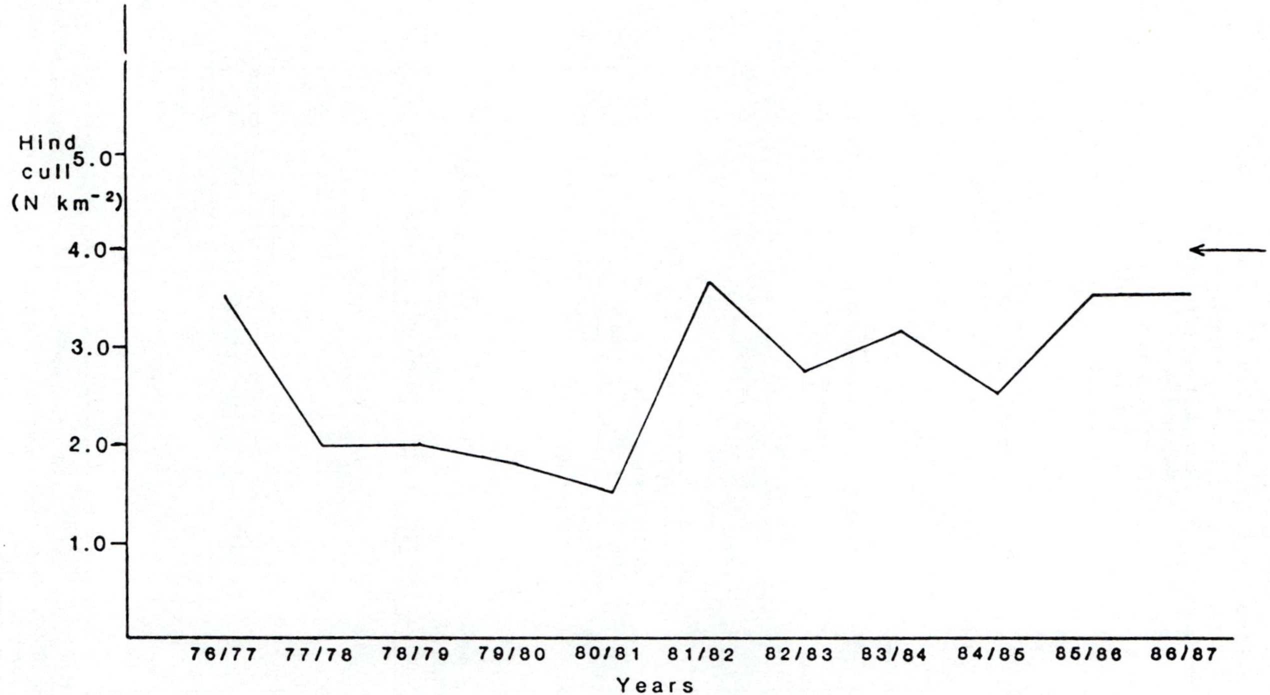

Figure 9.8 Red deer hind culling – Glencripesdale. This assumes low reproductive performance and high mortality. The cull requirements are approx. 4 km^{-2} based upon 40 deer km^{-2}.

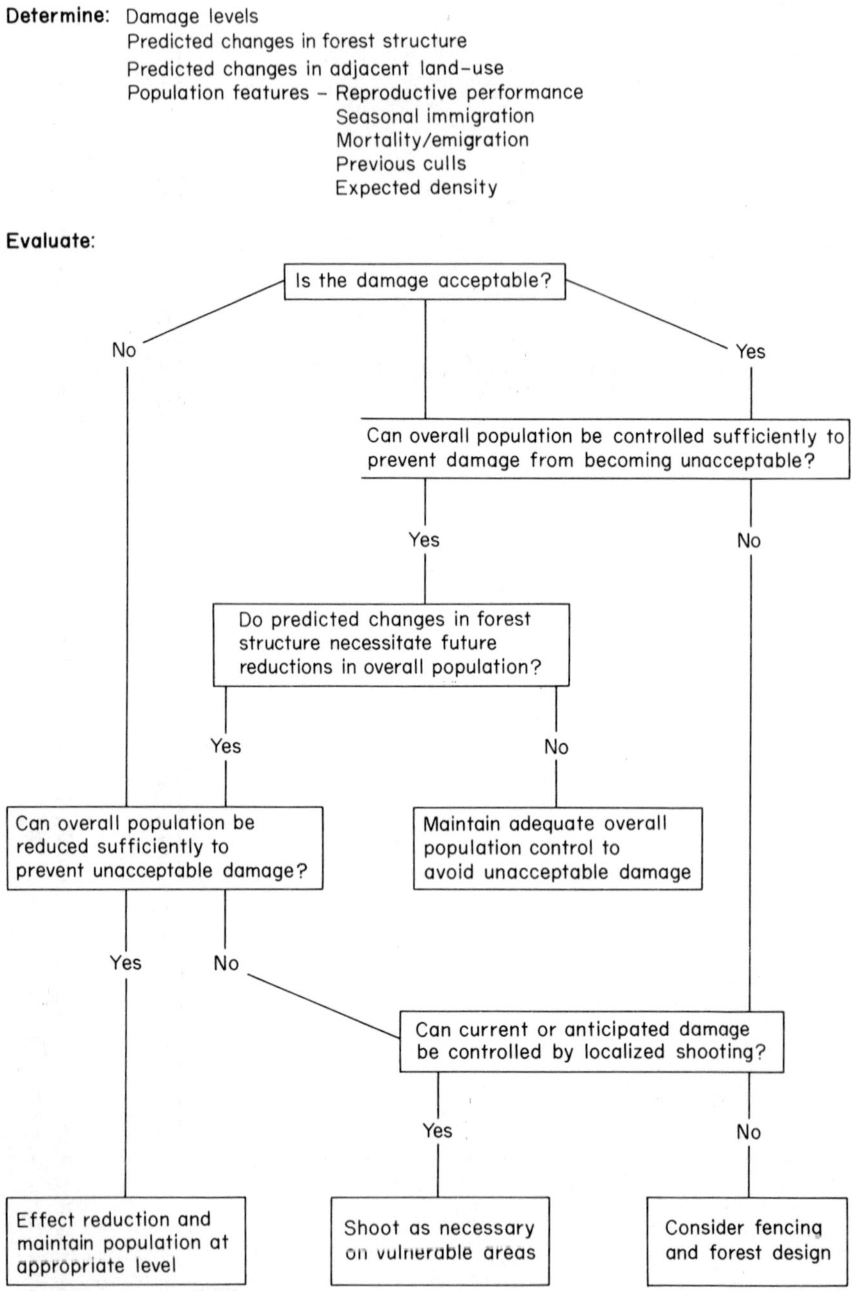

Figure 9.9 Red deer management strategy (after Ratcliffe, 1987c).

greater at high deer densities. Underexploitation may also be interpreted as a direct loss of revenue from venison or alternatively as increased capital (total population size) to be exploited at some future time. However, an inability to exercise control over current population levels suggests that this latter approach will only add to future problems including the risk to commercial tree species and sensitive plant communities and their dependent animals. An inability to achieve total population control does not mean that damage levels cannot be reduced by shooting deer on or near vulnerable areas. Managers must face the reality of attempting to achieve a balance of population control against damage control (Figure 9.9).

Whereas the above discussion focuses on red deer and illustrates the problems of control of deer populations using data for that species, recent work suggests that the population dynamics and densities of sika deer are very similar to those of red deer and the same general principles would apply (Ratcliffe, 1987b and unpublished).

THE COSTS OF CONTROL

Unlike the costs of damage, the costs of control measures can be readily calculated, though these costs may not always be justified. The value of control can only be measured against the benefits gained.

In commercial forests managed by the Forestry Commission in Scotland, deer culling is a very large part of rangers' duties. Additionally, a large length of fence is erected to protect plantations from access by red deer (Ratcliffe, 1984a). The costs of employing rangers to kill deer amounts to about £3 ha^{-1} $year^{-1}$ and the cost of erecting and maintaining fences, about £4 ha^{-1} $year^{-1}$. About 240 000 ha of Forestry Commission forests within the red deer range do not fence against deer, whereas some 300 000 ha employ both fencing and rangers. This represents a total annual cost of about £2.8 million to protect Forestry Commission forests from red deer. Currently we are unable to assess accurately the benefits which arise from this.

CONCLUSIONS

Damage to commercial tree crops, habitat impact, reduction in deer quality, and the continuing hybridization of red and sika deer all demand increased effort to effect control over population growth and increasing distribution.

Few, if any, populations of red and sika deer occupying commercial forests are being controlled, and populations are either continuing to increase as a result of improvements in range quality and habitat structure, or are regulating their densities at a high level by density-dependent mechanisms.

It is apparent that the constraints imposed by dense habitats and the legal methods and timing of control are preventing the effective control of red and sika deer populations. The creation of more open space in commercial forests will almost certainly make culling easier, but if felling creates an intimate mosaic of open areas and thickets this will improve the structure of the habitat for deer. Carefully designed forests and well-planned culling operations must complement one another.

REFERENCES

Burdekin, D.A. and Ratcliffe, P.R. (1985) Deer damage to Sitka – What does it really cost? *Forestry and British Timber,* August 1985.

Kampmann, H. (1983) Investigations on the effect of roe deer browsing of side shoots on the height growth of spruce plants. *Z. Jagdwissenschaft,* **29**, 235–43.

Lister, A.M. (1984) Evolutionary and ecological origins of British deer. *Proc. R. Soc. Edin.,* **82B**, 205–29.

Mitchell, B., Staines, B.W. and Welch, D. (1977) *Ecology of red deer. A research review relevant to their management in Scotland.* Institute of Terrestrial Ecology, Cambridge.

Ratcliffe, P.R. (1984a) Population dynamics of red deer (*Cervus elaphus*) in Scottish commercial forests. *Proc. R. Soc. Edin.,* **82B**, 291–302.

Ratcliffe, P.R. (1984b) Population density and reproduction of red deer in Scottish commercial forests. *Acta Zool. Fenn.,* **172**, 191–2.

Ratcliffe, P.R. (1987a) Red deer population changes and the independent assessment of population size. *Symp. Zool. Soc. Lond.,* **58**, 153–65.

Ratcliffe, P.R. (1987b) Distribution and current status of sika deer *Cervus nippon,* in Great Britain. *Mammal Rev.,* **17**, 39–58.

Ratcliffe, P.R. (1987c) The management of red deer in upland forests. *Forestry Commission Bulletin,* **71**, HMSO, London.

Ratcliffe, P.R. (1987d) The management of red deer in the commercial forests of Scotland related to population dynamics and habitat changes. Unpublished PhD Thesis. University of London.

Ratcliffe, P.R. (1988) The management of red deer populations resident in upland forests in *Wildlife Management in Forests* (ed. D. Jardine), Institute of Chartered Foresters, Edinburgh.

Red Deer Commission (1981) *Red Deer Management.* HMSO, London.

Ritchie, J. (1920) *The influence of man on animal life in Scotland: a study in faunal evolution.* Cambridge University Press, Cambridge.

Staines, B.W. and Ratcliffe, P.R. (1987) Estimating the abundance of red deer (*Cervus elaphus* L) and roe deer (*Capreolus capreolus* L) and their current status in Great Britain. *Symp. Zool. Soc. Lond.,* **58**, 131–52.

Staines, B.W. and Welch, D. (1984) Habitat selection and impact of red deer (*Cervus elaphus* L) and roe deer (*Capreolus capreolus* L) in a Sitka spruce plantation. *Proc. R. Soc. Edin.*, **82B**, 303–9.

Tee, L.A., Rowe, J.J. and Pepper, H.W. (1985) *Mammal and Bird Damage Questionnaire 1983. Forestry Commission Research and Development Paper* **137**. HMSO, London.

Wegge, P. (1973) Population tables for red deer. Unpublished paper Norwegian State Game Research Inst., Trondheim.

Whitehead, G.K. (1972) *Deer of the World.* Constable, London.

Ten

Deer and habitat relations in managed forests

M.J. HANNAN AND J. WHELAN

SUMMARY

Habitat utilization by deer in three coniferous woodland areas in Co. Wicklow, Ireland was investigated between February and November 1985 and damage to trees was assessed. Deer pellet groups were found in greater numbers in areas planted since 1968, indicating greater activity in those areas. Pellet group density was also found to be higher in areas of woodland close to thicket and with a higher degree of lateral cover.

Damage assessment was confined to areas planted since 1968. Browsing was the most common form of damage in two of the three study areas while bark stripping was most prevalent in the third. Fraying and thrashing were the least common forms of damage. Approximately 11% of trees examined in the Wicklow area had their leaders browsed whereas 20% had their side shoots browsed. The boles were stripped in 15% and the side shoots stripped in 7%. The percentage damage estimates were 41.3%, 51.5% and 35.2% for the three study areas.

The proximity of thicket and the newly planted areas appeared to be related to the incidence of damage ($p < 0.001$) with damage significantly lower within 50 m of thicket and significantly higher within 50 m of newly planted areas. Tree age was also a good predictor of damage occurrence.

INTRODUCTION

Intensive research on the utilization of vegetation by the Cervidae has been carried out in New Zealand (e.g. Holloway, 1950; Clarke, 1972), North America (e.g. Black, Dimock, Evans and Rochelle, 1979) and Europe (e.g. Larner, 1977; Bobek, Perzanowzki, Siwanowicz and Zielinski, 1979; Miller, Kinnaird and Cummins, 1982; Staines and Welch, 1984). Most damage work has been concerned with the description of the damage but there is little understanding of the principles and mechanisms involved (Miles and Kinnaird, 1979; Burdekin and Ratcliffe, 1985). The effects of the forest structures and management practices on the feeding behaviour of deer must be known before damage can be prevented or integrated control measures adopted (Gullion, 1982). In this chapter we examine the influence of some forest structures and habitat types on deer utilization, as reflected by pellet group counts we attempt to quantify the various types of damage inflicted on trees; finally, we examine the relationship between habitat use and damage as influenced by various parameters of forest structure.

STUDY AREAS

Three study areas for the habitat utilization and damage surveys were selected in Co. Wicklow (Figure 10.1). All three areas were owned and managed by the Forest Service, Department of Forestry, Dublin. The three areas comprised parts of Wicklow's largest forests. All were at high elevation (between 120 and 460 m) and established on peaty podsols, brown podsols and gleys; rainfall in the area varies from 1000 to 2000 mm per annum.

All study areas consisted mostly of Sitka spruce (*Picea sitchensis*) (70–75% of total). Some Douglas fir (*Pseudotsuga menziesii*) and Japanese larch (*Larix decidua*) have been recently planted whereas Scots pine (*Pinus sylvestris*) was present in the older stands. There were roughly equal numbers of stands in age-classes ranging from just planted to over seventy years old in each study area. The age-classes of each growing stock were intermixed to some extent in each forest block.

Ground cover in the study areas consisted of heather (*Calluna vulgaris*), whortleberry (*Vaccinium myrtilus*), purple moor grass (*Molinia caerulea*), *Erica* spp. and gorse (*Ulex europaeus*). Sika and fallow deer have been observed on all three study areas while red deer were observed in study areas 2 and 3. Occasional trespassing by domestic animals did occur in each study area.

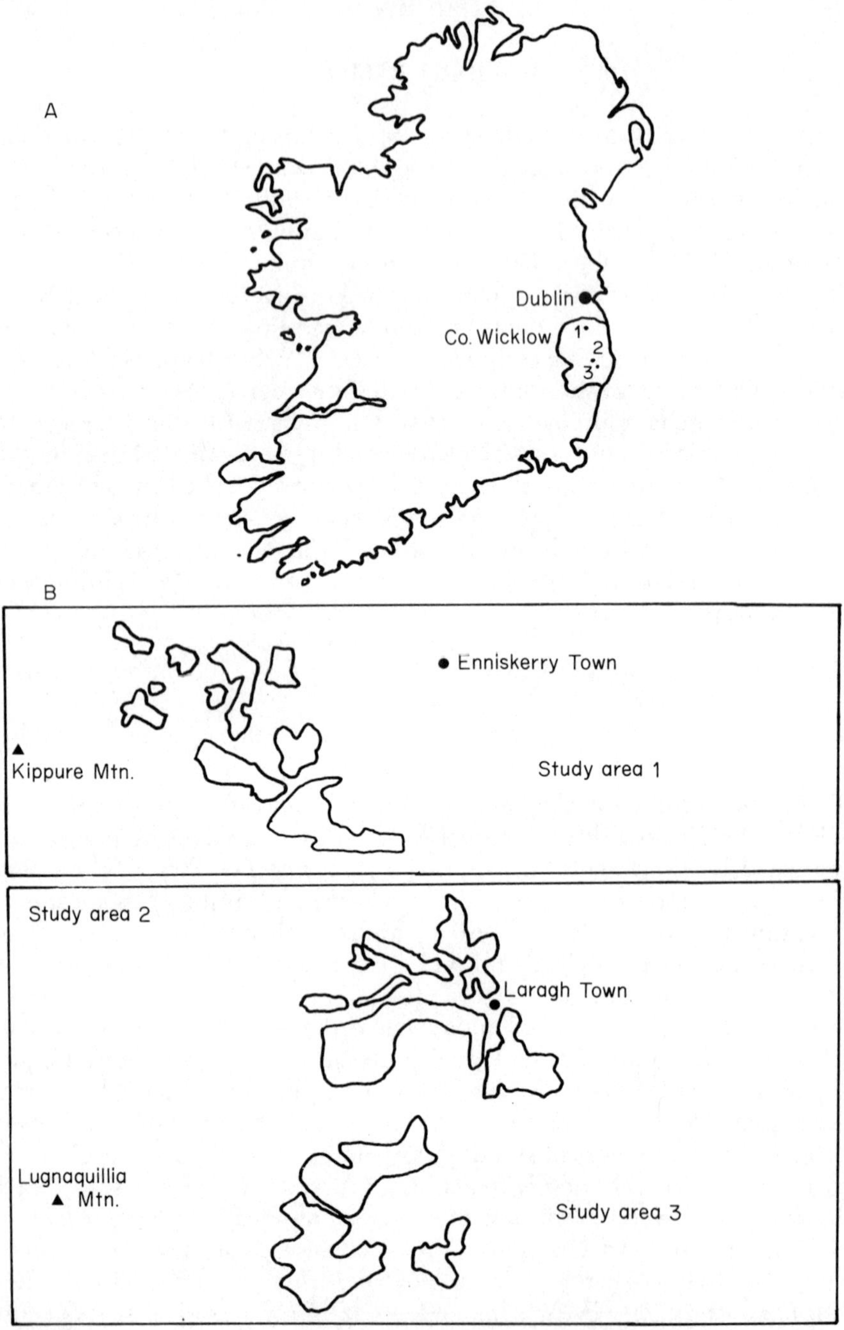

Figure 10.1 (a) The location of study areas in Co. Wicklow. Study area 1 – Glencree to Enniskerry, Study area 2 – Glendalough to Laragh, Study area 3 – Greenane to Glenmalure. (b) Outline map of the three study areas (Scale 1 : 158 400).

METHODS

Pellet group analysis

Habitat utilization by deer was gauged by pellet-group counts (Dasmann and Taber, 1955; Neff, 1968; Mitchell and McCowan, 1980; Mitchell, McCowan and Campbell, 1982; Putman, 1984). It was assumed that pellet-group density would be highest where deer spent most of their feeding time (Mitchell, McCowan and Campbell, 1982; Leopold, Krausman and Hervert 1974). In order to minimize the length of time taken to travel from one sampling unit to the next, relative to the length of time taken to sample each unit, large sampling units were used (Neff, 1968). Transects were used as sampling units as sampling efficiency on large circular plots is low (Smith, 1968). A pellet-group survey in Knockrath woodlands, Co. Wicklow revealed that the optimum transect length was 200 m (Hannan, 1984).

A total of 19 transects were established at random in each study area. Each transect was 200 m long. Pellet groups were cleared from these transects in the periods March–April and August–September, to a distance of 5 m on each side of the transect axis. Thereafter, transects were surveyed after 92 days, recleared and counted again after 49 days. To allow for error in relocation of the transect only pellet groups within 3 m of the axis were actually counted. At the same time vegetational characters around the transect were assessed from analysis of three 10 m^2 plots. Vegetation cover was estimated for:

Ground layer (less than 15 cm)
Field layer (15–60 cm)
Shrub layer (60–180 cm)
Canopy layer (greater than 180 cm)

Aspect and slope were also measured for each plot. Lateral cover was measured by pacing from a point in each plot where a white upright board measuring 190 by 25 cm was stationed. The number of paces to the point where vegetation obscured the board from view was the Lateral Cover Statistic. Data collected from each of the three plots were combined for each transect.

Along the transect the distance from open space, newly planted areas, thicket, water, mature timber, clearfelled areas, roads and ride-lines was estimated and assigned to four distance classes: 0–50 m, 51–100 m, 101–200 m, >200 m. All points along each transect were within the minimum and maximum distance of the class to which it was allocated.

Damage data collection

During August and September 1985 damage to trees in the three study areas was assessed. Damage was assessed as leading shoots browsed, side shoots browsed, bole stripped, branch stripped, tree frayed or thrashed, damage of economic significance and trees already dead; the number of trees in each damage category were estimated for each study area. Damage assessments were based on examination of clusters of trees in each study area. In each study area 30 points were selected at random and 20 trees around each centre-point taken as one cluster (Freeze, 1962). The number of damaged trees in each cluster was counted; a total number of 1800 trees being examined in this manner.

Clusters were allocated by using a grid superimposed onto the areas to be examined. From each cluster centre point the slope and aspect were measured. Distances to other habitat types and features were estimated in the manner described for pellet-group data collection. The age of trees in the cluster was estimated from whorl counts and age confirmed later with the local forest office.

DATA ANALYSIS

Pellet group data analysis

The differences between the mean number of pellet groups in study areas were tested for significance using one-way analysis of variance (Campbell, 1974). Analysis of variance was also used to test for differences in pellet-group density with aspect (north, south, east and west) as well as for differences with distance from habitat types and features. The influence of vegetation in each of four layers as well as of slope and lateral cover on pellet-groups was gauged by stepwise multiple regression (Weisberg, 1980). The stepwise procedure of the SAS: statistics package at University College Dublin was used. The FORWARD, BACKWARD and MAXR options were used for model development (SAS User's Guide: Statistics).

The Forward Selection procedure worked by adding to the model the independent variable that met three equivalent criteria (Weisberg, 1980).

1. It had the highest sample partial correlation with the dependent variable.
2. Adding the variable increased R^2 (the multiple correlation coefficient) more than any other single variable.
3. The variable had the largest F-statistic of any of the variables that were not in the model.

Conversely, the Backward Elimination method started with the full model and removed one variable at each step. The variable to be removed was chosen to be the one that had the smallest *F*-value of all the variables in the equation or that caused the smallest change in R^2 or had the smallest partial correlation with *Y* (adjusting for all the other variables left in the model). The MAXR option identified the best model when all variables were included. All data collected as percentages were transformed for analysis using Arcsin Transformation (Freeze, 1962).

Damage data analysis

Differences in the percentage damage in the different study areas, within study areas, and differences with distance from different habitat types were analysed with one-way analysis of variance, once again using transformed data. Stepwise regression was used to examine the influence of slope and age on damage.

RESULTS

Habitat utilization

Analysis of variance showed that the mean numbers of pellet groups observed on transects in each study area were not significantly different from each other at either the first ($F = 0.26$) or second ($F = 3.11$) sampling period. Pellet group numbers were generally highest in areas planted since 1968 when compared with areas planted between 1950 and 1968 and before 1950 in each study area ($F = 9.01$; $p < 0.001$; 1st sampling period: $F = 1.07$; NS; 2nd sampling period).

Of the habitat types and features examined, only the proximity of thicket appeared to exert any strong influence on numbers of pellet groups observed. The difference was significant during the first sampling visit ($F = 11.69$; $p < 0.001$; Table 10.1). Mean number of pellet-groups was also higher in the first two distance classes away from open space and water (0–100 m) than in areas further away. Pellet-group numbers were highest in areas more than 100 m from thinned stands suggesting that thinned stands were avoided by deer.

North- and south-facing slopes were utilized more than slopes facing east or west but differences recorded were not statistically significant. With data collected during the first sampling period, lateral cover was selected first into the multiple regression model at the 0.05 significance level, with slope and canopy vegetation subsequently included in addition. During the background elimination procedure, all variables in

Table 10.1 The mean number of pellet-groups observed on transects in categories of distance from thicket

Distance class	n	$X \pm 95\%$ Conf. limits	F	Signif. level
First sampling period (July/August 1985)				
0–50 m	32	15.28 ± 3.26		
51–100 m	5	2.20 ± 1.84	11.69	$p < 0.001$
101–200 m	8	5.13 ± 3.50		
>200 m	12	4.00 ± 1.69		
2nd sampling period (October/November 1985)				
0–50 m	27	5.48 ± 2.09		
51–100 m	1	–	0.90	N.S.
101–200 m	8	3.25 ± 1.77		
>200 m	16	4.25 ± 2.05		

the model were significant at the 0.05 level after the shrub, ground and field vegetation variables were removed. The predictive equation of the remaining model was:

$$Y = 0.16X_1 - 0.16X_2 + 0.35X_3 + 6.66$$

where X_1 = canopy vegetation; X_2 = lateral cover; and X_3 = slope.

The best six-variable model found when R^2 was maximized ($R^2 = 0.46$) was

$$Y = 0.42X_1 - 0.05X_2 + 0.02X_3 + 0.12X_4 - 0.18X_5 + 0.36X_6 + 9.98.$$

where X_1 = ground vegetation; X_2 = field vegetation; X_3 = shrub vegetation; X_4 = canopy vegetation; X_5 = lateral cover and X_6 = slope.

Analysis of data from the second sampling period also showed a relationship between the pellet-group density and lateral cover, but with this latter data set no relationships approached statistical significance at $p = 0.05$.

Deer damage

Browsing was by far the most common form of damage inflicted on trees in Co. Wicklow (Table 10.2). Browsing on side-shoots was more common than browsing on leading shoots. Between the study areas, browsing was more common in study areas 1 and 2 whereas stripping was more common in study area 3. Fraying and thrashing were by far the least common forms of damage. In general bole stripping was more prevalent than branch stripping. Of the 1800 trees examined, however, only 25 were likely to die directly as a result of damage inflicted by deer.

Table 10.2 The number of trees in categories of damage type and severity in each study area in Wicklow (600 trees were assessed in each area).

Damage category	Study area 1	2	3
Leading-shoot browsed	87	90	38
Side-shoots browsed	114	201	70
Bole stripped	73	93	115
Branch stripped	38	40	60
Tree frayed/thrashed	0	4	4
Damage of economic significance	9	2	14
Tree already dead	31	21	53

In Wicklow less damage was inflicted on plots consisting of Sitka spruce alone than on plots consisting of other species or combinations of species. The difference was not significant, however, probably because the variability within the population was too high.

Damage to trees was significantly lower ($F = 8.07$; $p < 0.001$) in areas within 50 m of thicket than in areas more than 50 m away (Table 10.3). Damage was significantly greater ($F = 21.92$; $p < 0.001$) in areas within 50 m of newly planted areas than in areas more than 50 m away (Table 10.4). These two habitat types appeared to exert the most influence on

Table 10.3 Mean damage in classes of distance from thicket of the three study sites

Distance class	n	$X \pm 95\%$ Conf. limits	F	Signif. level
0–50 m	46	29.96 ± 5.71		
51–100 m	4	53.37 ± 39.07	8.07	$p < 0.001$
101–200 m	8	45.97 ± 22.20		
>200 m	32	51.99 ± 7.39		

Table 10.4 Mean damage in classes of distance from newly planted areas of the three study sites

Distance class	n	$X \pm 95\%$ Conf. limits	F	Signif. level
0–50 m	51	47.17 ± 5.81		
51–100 m	0	—	21.92	$p < 0.001$
101–200 m	16	23.17 ± 5.93		
>200 m	23	26.17 ± 8.09		

Table 10.5 Mean damage in classes of distance from roads of the three study sites

Distance class	n	$X \pm 95\%$ Conf. limits	F	Signif. level
0–50 m	27	34.08 ± 8.02		
51–100 m	12	26.26 ± 11.71	6.82	$p < 0.001$
101–200 m	25	38.54 ± 7.99		
>200 m	26	54.79 ± 9.67		

damage incidence. Of the remaining habitat types and features, only roads appeared to be influential. Damage to trees increased significantly with increasing distance from roads ($F = 6.82$; $p < 0.001$, Table 10.5). Aspect did not appear to influence the incidence of damage.

During forward selection with the model development for regression of age and slope on damage only the relationship with age was selected at $p < 0.05$. During backward elimination, slope was removed at both the 0.1 and 0.5 level resulting in the equation

$$Y = -4.25X + 74.79$$

where X = age.

DISCUSSION

This chapter reports on habitat use and damage to trees in Co. Wicklow, Ireland. In general vegetation in the ground and field layers were poor predictors of deer utilization; pellet-group densities were primarily related to canopy vegetation, lateral cover and slope. Pellet-group counts suggest that thickets rather than newly planted areas were attractive regions within the portion of forest planted since 1968. In contrast, areas at or within 50 m of thicketed areas suffered significantly less damage than areas further away, while areas at or within 50 m of new plantations had significantly greater damage. This finding corresponds with others who have shown that browsing damage occurs while the trees are young and before they have entered the thicket stage of development (Maxwell, 1968; Black *et al.*, 1979 and Welch *et al.*, 1982), but that thicket areas are widely used for cover and shelter.

In general, the activity of deer was lower where the density of roads, ridelines and fire breaks was high (Hannan, 1984). Damage increased with increasing distance from roads although roads did not appear to influence pellet-group numbers. Lyon (1979) and Rost and Bailey (1979) also found that roads through forests were avoided by deer and in the latter case, roads prevented deer from utilizing highly nutritious feeding areas.

As already noted, age was a good predictor of damage. Higher incidences of damage were apparent in clusters where trees were youngest. This may help to explain why the incidence of leader-shoot browsing was high in comparison to side-shoot browsing considering the relative availability of each type to deer. Welch *et al.* (1988) also found that roe deer damaged proportionately more leaders than lateral shoots and cited the protection of lateral shoots by ground vegetation as a possible reason for this.

Mitchell and McCowan (1986) reported that deer damage and site occupancy by deer were positively related but not always significantly so. In the current study, it could only be shown that deer activity was greatest in areas within the study sites planted since 1968. Within those areas deer appeared to utilize thicketed areas where lateral cover was greatest but emerged from this cover to damage trees in newly planted areas. The regression models showed that lateral cover was a strong predictor of where pellet-group numbers were highest and that age was a good predictor of where damage was most inflicted. It has been widely accepted that deer of the genus *Cervus* are ecotonal (Mitchell, Staines and Welch, 1977; Clarke and Gilbert, 1982 and Bobek, Boyce and Kosobucka, 1984). It would be interesting to examine further the relationship between thicket and newly planted areas with a view to informing forest managers how best to intersperse the two. The size of areas where damage occurs should also be examined. Such information would be useful to foresters planning on where to plant.

ACKNOWLEDGEMENTS

We are indebted to the Forest and Wildlife Service who permitted us to carry out all field work on their lands. We acknowledge with thanks the help given to us by the foresters of Enniskerry, Glencree, Glendalough, Glenmalure, Greenane and Laragh Forests. We thank our colleagues in Agricultural Zoology for their support and advice.

REFERENCES

Black, H.C., Dimock, E.J., Evans, J. and Rochelle, J.A. (1979) Animal damage to coniferous plantations in Oregon and Washington. Part 1. A Survey 1963–1975. *Oregon State University Research Bulletin,* **25**, 45pp.

Bobek, B., Boyce, M. and Kosobucka, M. (1984) Factors affecting population density in south-eastern Poland. *J. Appl. Ecol.,* **21**, 881–90.

Bobek, B., Perzanowski, K., Siwanowicz, J. and Zielinski, J. (1979) Deer pressure on forage in a deciduous forest. *Oikos*, **32**, 373–80.
Burdekin, D.A. and Ratcliffe, P.R. (1985) Deer damage to Sitka – what does it really cost? *Forest. Br. Timber*, **14**, 26–7.
Campbell, R.A. (1974) *Statistics for Biologists*, 2nd ed., Cambridge University Press, Cambridge.
Clarke, C.M.H. (1972) Red deer in Northern South Island region: Their early impact. *N. Z. J. Forestry*, **17**, 37–42.
Clarke, T. and Gilbert, F.F. (1982) Ecotones as measures of deer habitat quality in Central Ontario. *J. Appl. Ecol.*, **19**, 751–8.
Dasmann, R.F. and Taber, R.D. (1955) A comparison of four deer census methods. *California Fish and Game*, **41**, 225–8.
Freeze, F. (1962) Elementary forest sampling. *Agricultural Handbook no.* 232. USDA Forest Service.
Gullion, G.W. (1982) Forest-widlife interactions in *Introduction to Forest Science* (ed. R.A. Young). Wiley, New York.
Hannan, M.J. (1984) *A cost benefit analysis of deer on Knockrath Woodland.* Unpublished B.Agr.Sc. Thesis, University College Dublin.
Holloway, J.T. (1950) Deer and the forests of western Southland. *N. Z. J. Forestry*, **6**, 123–37.
Larner, J.B. (1977) Sika deer damage to mature woodlands of South-western Ireland. *Proceedings of the XIII Congress of Game Biologists*, pp. 192–202.
Leopold, B.D., Krausman, P.R. and Hervert, J.J. (1984) The pellet-group census technique as an indicator of relative habitat use. *Wildl. Soc. Bull.*, **12**, 325–6.
Lyon, L.J. (1979) Habitat effectiveness for elk as influenced by roads and cover. *J. Forestry*, 77, 658–60.
Maxwell, H.A. (1968) Red deer and forestry practice. *Scottish Forestry*, **22**, 18–23.
Miles, J. and Kinnaird, J.W. (1979) Grazing: with particular reference to birch, juniper and scots pine in Scottish Highlands. *Scottish Forestry*, **33**, 280–9.
Miller, G.R., Kinnaird, J.W. and Cummins, R.P. (1982) Liability of saplings to browsing on red deer range in the Scottish Highlands. *J. Appl. Ecol.*, **19**, 941–51.
Mitchell, B. and McCowan, D. (1980) Estimating and comparing densities of red deer, *Cervus elaphus* L., in concealed habitats. *Institute of Terrestrial Ecology, Annual Report 1979*, pp. 7–13. Institute of Terrestrial Ecology, Cambridge.
Mitchell, B. and McCowan, D. (1986) Patterns of damage in relation to the site preference of deer in an enclosed plantation of sitka spruce and lodgepole pine. *Scottish Forestry*, **40**, 107–17.
Mitchell, B., McCowan, D. and Campbell, D. (1982) Faecal depositions as indicators of site use by red deer. *Annual Report. Institute of Terrestrial Ecology*, pp. 85–7. Institute of Terrestrial Ecology, Cambridge.
Mitchell, B., Staines, B.W. and Welch, D. (1977) *Ecology of Red Deer. A Research Review Relevant to their Management in Scotland.* Institute of Terrestrial Ecology, Cambridge.
Neff, D.J. (1968) The pellet-group count technique for big game trend, census and distribution: A Review. *J. Wildl. Mgmt*, **32**, 597–614.
Putman, R.J. (1984) Facts from faeces. *Mammal Rev.*, **14**, 79–97.
Rost, G.R. and Bailey, J.A. (1979) Distribution of mule deer and elk in relation to roads. *J. Wildl. Mgmt*, **43**, 634–41.
SAS Users Guide: Statistics (1982) S.A.S. Institute, Cary, North Carolina.
Smith, R.H. (1968) A comparison of several sizes of circular plots for estimating deer pellet group density. *J. Wildl. Mgmt*, **32**, 585–91.

Staines, B.W. and Welch, D. (1984) Habitat selection and impact of red and roe deer in a sitka spruce plantation. *Proc. R. Soc. Edin.* **82B**, 303–19.

Weisberg, S. (1980) *Applied Linear Regression*. Wiley, New York.

Welch, D., Chambers, M.G., Scott, D. and Staines, B.W. (1988). Roe-deer browsing on spring-flush growths of sitka spruce. *Scottish Forestry*, **41**, 33–43.

Welch, D., Staines, B.W., Scott, D. and Catt, D.C. (1982) Browsing damage done by red deer to stands of sitka spruce. *Annual Report. Institute of Terrestrial Ecology* pp. 10–12. Institute of Terrestrial Ecology, Cambridge.

Eleven

Impact of red and roe deer on Scottish woodlands

B.W. STAINES AND D. WELCH

Populations of red (*Cervus elaphus*) and roe (*Capreolus capreolus*) deer are increasing throughout northern Europe. Estimates of the total red deer population in Scotland alone are 290 000 for deer living on open moorland, with a possible 25 000–50 000 in forestry plantations (Staines and Ratcliffe, 1987).

In Scotland, both species of deer can cause considerable damage to native and planted woodlands and may affect regeneration and composition (Mitchell, Staines and Welch, 1977). The main forms of damage include browsing the leading and lateral shoots, eating the bark (red deer) and fraying or rubbing the stems with the antlers. The first two are considered the most important.

The relative susceptibilities of different trees to these forms of deer damage vary (Mitchell, Staines and Welch, 1977). Of the broadleaves, willows, rowans and aspens are highly susceptible to browsing and birch and alder least affected. Among the conifers, Norway spruce and Douglas fir are more vulnerable than Scots pine and Sitka spruce. Likewise Norway spruce and lodgepole pine are very susceptible to barking, whereas Sitka spruce and birch are least affected. Even within species differences occur, and thin-barked provenances of lodgepole pine (generally of inland origin) are more heavily damaged than the thicker-barked coastal provenances. Such differences in susceptibility are, however, relative, and if nothing else is available *any* species can be severely damaged.

Recent work on red and roe deer in forestry plantations by the Institute of Terrestrial Ecology, the Forestry Commission and the Red Deer Commission has involved studies on the population dynamics, ranging and feeding behaviour of the deer and quantitative assessments of the effects of different forms of damage on the trees (e.g. Staines and Welch, 1984; Ratcliffe, 1984; Ratcliffe and Staines, 1982).

Rides and open areas close to forest cover are highly favoured by both species, and the bare intermediate-age stages of the forest are least favoured (Staines and Welch, 1984). The early stages of the forest are heavily utilized, with roe deer tending to use the pre-thickets more than red deer. Conversely red deer are more common than roe in the thicket stages, especially when these contain many open areas. Red deer are primarily grazers with grasses being the principal foods in summer and, when available, in winter (Staines and Welch, 1984). Roe deer eat mainly forbs in summer and both deer species feed much on heather (*Calluna vulgaris*) in winter. Conifer browse is eaten mostly (1) in a short period in spring at bud burst and (2) in late winter. In winter roe deer eat relatively more conifer shoots than do red deer.

Observations on the trees themselves confirm the findings of rumen analysis, there being two peaks in damage in winter and spring. The incidence of leader browsing on Sitka spruce (*Picea sitchensis*) peaks in the third year after planting when often half the trees are damaged. Once trees exceed a height of 80 cm their leaders are rarely browsed; trees usually suffer leader browsing two to three times before reaching this height and very few escape. Browsing apparently has little effect on vertical growth on good sites, although on some poor areas trees may need protection for up to 5 years in order to reach a safe height. However, after leader browsing many trees become multi-stemmed, which can adversely affect the value of the timber crop (Staines and Welch, 1984; Welch *et al.*, 1983).

Sitka spruce is less likely to be bark-stripped than many other species. Average rates are less than 1% per year of the susceptible age-classes of trees (8 years to 40 years), although local ‘hot-spots’ of heavier damage occur. When stands first become vulnerable to bark-stripping, the trees damaged are the larger ones; after a few years the trees barked are of average size, whilst in older stands the smaller, weaker trees are most affected, many of which would soon die anyway. Hence mere estimation of the proportions of trees damaged in particular stands gives an inadequate assessment of the actual economic effects on the final crop.

The sizes of wounds caused by bark-stripping are smaller on Sitka than, for example, on Norway spruce. Consequently, after wounding Sitka is less affected by secondary damage from fungal rots and staining. However, callousing and resin-inclusion into the wood may result in a significant de-grading of the timber.

Long-term studies on the effects of deer damage are continuing so that the economic costs of damage and deer control can be fully evaluated.

REFERENCES

Mitchell, B., Staines, B.W. and Welch, D. (1977) *Ecology of Red Deer: A Research Review Relevant to their Management in Scotland.* Institute of Terrestrial Ecology, Cambridge, 74pp.

Ratcliffe, P.R. (1984) Population dynamics of red deer (*Cervus elaphus* L.) in Scottish commercial forests. *Proc. R. Soc. Edin.*, **82B**, 291–302.

Ratcliffe, P.R. and Staines, B.W. (1982) Red deer in woodlands: research findings in *Roe and Red Deer in British Forestry*, British Deer Society, Warminster, pp. 42–53.

Staines, B.W. and Ratcliffe, P.R. (1987) Estimating the abundance of red deer (*Cervus elaphus* L.) and roe deer (*Capreolus capreolus* L.) and their current status in Great Britain. *Symp. Zool. Soc. Lond.*, **58**, pp. 131–52.

Staines, B.W. and Welch, D. (1984) Habitat selection and impact of red (*Cervus elaphus* L.) and roe (*Capreolus capreolus* L.) deer in a Sitka spruce plantation. *Proc. R. Soc. Edin.*, **82B**, 303–19.

Welch, D., Staines, B.W., Scott, D. and Catt, D.C. (1983) Browsing damage done by red deer to young stands of Sitka spruce. *Ann. Rept. Inst. Terr. Ecol.*, 1982, pp. 10–12.

Editor's note: This paper is presented here only in summary. The full text will be found in Staines, B.W. and Welch, D. (1989) An appraisal of deer damage in conifer plantations in *Forestry and Deer* (Proceeding of discussion meeting, June 1987; ed. R. McIntosh), Institute of Chartered Foresters, Edinburgh (in press).

Twelve

Demographic implications for the control of grey squirrels

J. GURNELL

SUMMARY

This chapter is in three parts. First, the demographic characteristics of grey squirrels are reviewed. In stable forest habitats, known numbers of squirrels exhibit marked annual cycles around long-term averages which are characteristic for particular types of forest habitat. The level of average abundance varies according to the age and tree species composition of the forest and the forest's contingent long-term average supplies of squirrel foods (e.g. tree seeds and caterpillars). Changes in numbers between seasons can be large. Food shortage and the severity of winter weather are believed to be the key limiting factors which determine abundance during the annual cycle but there are inherent difficulties in quantifying them. Control has only a short-term effect on population dynamics.

Secondly, against the above information the temporal and spatial aspects of squirrel damage are considered. In Britain nearly all economic damage occurs between June and July. The spatial relationships among different types of forest, in particular, potentially high density and damage-vulnerable habitats, can be important with respect to squirrel distribution during the damage period. It is concluded that control measures need not be implemented every year and when necessary they need only be carried out prior to and during the damage period and, where possible, only in potentially high density habitats.

Thirdly, a simple index method of predicting relative abundance during the damage period is considered. This is based on the positive effects of winter food abundance on squirrel demography and its inverse effects on squirrel trappability. The decision risk of using such an index method as a management tool is discussed.

INTRODUCTION

In Britain, both grey and red squirrels remove bark from trees, sometimes on a scale which causes an economic loss of timber, particularly in plantations between 10 and 40 years old (Kenward, 1983; Kenward and Parish, 1986; Gurnell and Pepper, 1988; and Chapter 13). This chapter concentrates on the grey squirrel. The principal method of controlling grey squirrels in areas where red squirrels are not at risk is by using a 0.02% concentration of the anticoagulant poison warfarin on a whole wheat bait placed in specially designed hoppers in and around crops vulnerable to damage. The provisions governing this method of control are laid down by the 1973 Squirrels (Warfarin) Order under the 1972 Agricultural Act (Rowe, 1980) and the 1981 Wildlife and Countryside Act. A cost-effective control programme, which will have the added benefit of minimizing the annual large-scale use of poison, will depend on predicting where and when damage will occur and when and where squirrel populations should be controlled. Such a prediction assessment must involve, on the one hand, an understanding of the demographic and spatial dynamics of squirrels, and on the other, the factors which characterize types of forest as being particularly vulnerable to damage; such forest habitats are referred to as damage vulnerable habitats (Gurnell and Pepper, 1988).

This chapter considers the population demography of grey squirrels in British forests as it relates to the timing of damage (primarily June and July) and the types of habitat most vulnerable. These habitats consist mainly of 10- to 40-year-old broadleaf trees such as beech, oak and sycamore. There is increasing concern about grey squirrel damage to coniferous trees in the Midlands, North Yorkshire and Wales but little is known about the dynamics of squirrels in such habitats. Finally, a simple field method for predicting future squirrel abundance is described and its utility as a management tool is assessed.

DEMOGRAPHY AND SPATIAL DYNAMICS

In stable forest habitats, known numbers of squirrels exhibit marked annual cycles around long-term averages characteristic of those habitats (Gurnell, 1983, 1987). Closely linked with the demography of squirrels within specific habitats is their spatial interrelationships (e.g. foraging movements, dispersal) among neighbouring habitats; these are referred to as the spatial dynamics of the animals. Relevant aspects of average abundance, spatial dynamics and annual cycles are briefly discussed below; a more wide-ranging discussion of these topics can be found elsewhere (Gurnell, 1987). At the outset, however, one factor which affects the collection of data about squirrel populations should be noted: squirrels are not attracted to bait at traps, or indeed poison hoppers, when favoured foods such as tree seeds are abundant (Dubock, 1979; Gurnell, 1983). As a direct result of this, quantitative information concerning the demography of autumn and winter populations is frequently poor as a result of reduced trappability, and the most reliable available data concern spring and summer populations. In consequence, much of the analysis to follow refers to spring and summer populations.

Long-term average abundance

It is believed that the level of average abundance varies according to the type of forest (e.g. tree species, age) and its contingent long-term average supplies of squirrel foods (e.g. tree seed, caterpillars). However, in spite of several past studies on the population dynamics of grey squirrels, little is known about long-term average abundances in many types of forest, especially coniferous forests. Densities of grey squirrels in broadleaved forests range from 3 to 10 squirrels ha^{-1} or more (Gurnell, 1983). In contrast, densities of grey squirrels in Scots and Corsican pine plantations in Thetford Forest appear to be considerably less than one squirrel ha^{-1} (Gurnell and Taylor, unpublished), but Thetford Forest may not be considered typical of coniferous forests elsewhere in the country. High-density or potentially high-density habitats such as those consisting of large-seeded broadleaved trees, are significant in not only providing a source of squirrels for moving into other habitats but also a focus for control operations. In the latter instance, and as discussed further below, the squirrels removed will be replaced by squirrels drawn in from surrounding habitats.

Importantly, squirrel densities in most habitats considered particularly vulnerable to damage, i.e. those consisting of pure stands of broadleaved trees, are relatively low, e.g. less than one squirrel ha^{-1} (e.g. MacKinnon, 1976), at least until the trees reach 30–40 years old. However, sometimes broadleaved trees such as beech are planted in intimate mixtures with a nurse crop of spruce or pine. Here, when the coniferous trees reach 25–30 years old and start to cone, conditions may be such as to support at least a moderate density of squirrels. Specific demographic studies need to be carried out in these sorts of habitat.

Spatial dynamics

Equally important to our understanding of the factors which affect long-term average abundance are the spatial dynamics of squirrels in mixed forests which consist of different species/age blocks of trees. Little is known about these dynamics and how they are affected by seasonal changes in the carrying capacities of the constituent blocks or habitat patches (Gurnell, 1981). Of particular concern is whether habitats, especially potentially high density habitats, build up high numbers relative to their carrying capacities during May to July. If this occurs, there is a potential surplus of squirrels which may move among forest blocks at this critical time. Thus, damage to pure stands of young oak or beech trees will result mainly from squirrels visiting these areas from, for example, high density habitats (e.g. Don, 1981).

Two different types of movement affect the changing pattern of distribution of squirrels throughout the year. First, squirrels can range over large areas in the course of their normal activities of feeding and breeding (i.e. home range). Range areas of between 0.5 ha and 10 or more hectares have been recorded and are influenced by habitat type, food supplies, season and population density (Gurnell, 1987). Both a change in diet from high-energy tree seeds during the autumn and winter to low-energy flowers and buds in the spring, and breeding activity of males during the spring and summer can result in increased range sizes before and during the damage period, June and July (Gurnell, 1987). More important than home range size *per se* is that squirrels regularly move between different habitats. In a study in southern England more than 30% of the squirrels were found to commute regularly between a 9 ha oak wood and surrounding habitats (see Ludolf, 1986) and squirrels have been known to use areas as much as a kilometre apart (Don, 1981; Ludolf, 1986). Under these circumstances and particularly where forests consist of small blocks of trees (e.g. 5 ha to 40 ha) it becomes difficult to define a population in both time and space.

Secondly, two main periods of squirrel dispersal have been identified:

(1) during the spring and/or summer; this dispersal phase is particularly associated with movements of young of the previous year which may be returning from marginal, overwintering habitats to try and establish themselves in breeding populations, and (2) during the autumn, with in this case many dispersal movements by young born earlier in the year. The amount of autumn dispersal is particularly influenced by autumn tree seed supplies. The home range of squirrels seems to increase quite substantially prior to dispersal (Kenward, 1985) but few data exist to indicate how far dispersing squirrels may go. Certainly, some move distances of more than 1 km (Gull, 1977) and there are records of movements >3 km (Taylor, 1977). Dispersal and foraging distances are important in relation to the distance between potentially high density and damage-vulnerable habitats; these may be particularly relevant to blocks of trees in agricultural areas.

Annual cycles

This section illustrates the salient features of annual cycles by reference to data collected so far in an ongoing 12-year study of the demography of grey squirrels in a 9 ha oak wood in southern England; a detailed analysis of these data will be presented elsewhere, but data from some of the early years of the study have been discussed by Gurnell (1981, 1983).

Population trends. The 160-year-old oak wood had a high average density of squirrels over the study period of approximately seven squirrels ha^{-1}. Summer numbers of squirrels known to be present ranged from 47 squirrels to 88 squirrels with the exception of two years. In 1977, numbers reached 164 squirrels and this has been attributed to the one very large seed crop to occur during the study, in the autumn of 1976 (Gurnell, 1983). In 1982 numbers fell to 26 squirrels following a poor seed crop and a very cold spell the previous December (Gurnell, 1987). The male:female adult sex ratio over the period of study was 1.11.

Squirrel numbers fluctuated considerably during the year and there appeared to be no association between numbers of squirrels present in mid-winter and those present in June and July, the height of the damage period (see Gurnell and Pepper, 1988). Food limitation, particularly tree seeds, was a critical factor which influenced these fluctuations (Table 12.1). Further, the number of adult females present in the summer population appeared to inversely influence the change in population numbers through to the next summer. On the other hand, adult male numbers, which had a higher turnover of numbers between summers (3.75 squirrels entering or leaving the population for every resident squirrel), than females (1.75 entering or leaving the population for every

Table 12.1 Some regression analyses for the prediction of population change between summers for a 9 ha oak wood in southern England. The dependent variable is the change in squirrel population numbers from one summer to the next (PC). The independent variables are an index of autumn seed availability (FOOD), an index of the severity of winter weather (COLD) and the number of females present the first summer (FEMALES). PC = number of individuals present in first summer minus the number present in second summer. FOOD = log (seed biomass [g] 10 m^{-2}). COLD is the logarithm of the cold weather index as used by Gosling (1981); the cold weather index is calculated as the sum of the squares of the runs of days with minimum temperature ⩽0°C and maximum temperatures ⩽5°C. The regression model is of the form $y = a + b_1 x_1 + b_2 x_2 + \ldots + b_i x_i$, where i is the ith independent variable (Gurnell, unpublished).

Regression equation	R^2(%)	Significance of coeffs a	b_1	b_2	b_3	F
PC = 16.2 − 7.6 COLD	<1	ns	ns			<1
PC = 34.4 FOOD − 18.1	44	ns	**			6.96**
PC = 85.6 − 3.0 FEMALES	31	ns	ns			3.95
PC = 38.7 FOOD − 33.1 COLD − 55.4	51	ns	*	ns		4.16
PC = 35.1 FOOD − 3.1 FEMALES − 70.9	76	ns	**	**		12.61*

* $p<0.05$ ** $p<0.01$ ns not significant.

resident squirrel), were not related in any predictable way to population change between summers (or indeed to survival and recruitment). Little association between the severity of winter weather and population demography has so far been found (Table 12.1). One possible reason for this is that the effects of cold weather are influenced by food supplies. When food is plentiful, squirrel numbers at this site do not appear to be affected by cold weather unless unusually severe and long-lasting. (Further discussion on the interrelationships between food, weather and squirrel energetics can be found in Reynolds, 1985.) Moreover, severe cold weather late in the winter will have a greater impact than that towards the beginning of the winter.

Residency and immigration. Most losses of individuals from the population occurred between summer and the following spring (Table 12.2); it is believed that these resulted mainly from dispersal during the autumn and overwinter mortality. On average more females (56%) than males (39%) remained in the population between summers. Only about 25% of the juveniles remained in the population over their first winter.

Of the new adults which immigrated into the population between summers, about 70% actually entered the population between spring and summer (Table 12.2). Thus large numbers of adults moved into the population during the damage period which supports the notion that the

Table 12.2 Mean residency of adults and juveniles (expressed as percentage surviving from one summer to the next) and mean immigration rate of new adults (expressed as percentage of new adults of all adults present in second summer population) between summers. Data collected from a 9 ha oak wood in southern England, 1976–1987.

Period	Mean residency (%)				Mean immigration rate (% summer adult populn)	
	Males		Females			
	adults	juvs	adults	juvs	males	females
Summer to spring	43	23	60	30	16	12
Summer to summer	39	20	56	28	57	32

Source: Gurnell, unpublished data.

removal of squirrels from potentially high-density habitats at this time will draw squirrels away from damage-vulnerable habitats. The number of adult females entering the summer population was associated with total summer population numbers ($r = 0.77$, $n = 12$, $p < 0.01$); this was not so for adult males ($r = 0.53$, $n = 12$, $p > 0.05$). In summary, the spring to summer period is an important time as some squirrels move between areas and attempt to settle in high-density habitats.

Juvenile spring recruitment. When food supplies were good, the breeding season started early (December–January) and young were born in the spring (Table 12.3). In these years a second pulse of reproduction occurred with spring mating and summer litters (except 1977 when very high densities were thought to have suppressed litter production during the summer, see Gurnell, 1983). In contrast, in years when there were no autumn seed crops, the start of breeding was deferred until the spring and

Table 12.3 Relationship between seed crop size, the mean number of spring-born young recruited into the summer population and two measures of winter (January) trappability: the mean number of captures per trap-round (number of traps × number of rounds) and the mean number of captures per individual per trap period. Figures are average values from 12 years of study, 1976–1987 (Gurnell, unpublished). Multicapture traps were prebaited 7 days before they were set and trapping was carried out for 4 days, with a morning and evening inspection round of the traps on the first 3 days. For measurement of seed supplies, see Gurnell (1981).

Relative seed crop size	Mean capts. indiv.$^{-1}$ (n, sd)	Mean capts. trap-round^{-1}	Mean number spring-born young recruited into summer population (no. y^{-1})
None	1.76 (202, 1.028)	0.58	0
Poor	1.53 (99, 1.072)	0.47	14
Good	1.06 (18, 0.236)	0.06	43

no juveniles were present during June and July. Overall, the number of juvenile recruits in the summer population was moderately associated with the previous autumn seed supplies ($r = 0.71$, $n = 11$, $p < 0.01$).

Population model. This study suggests that a reasonable explanatory model of population fluctuations might eventually emerge from further research. However, from the point of view of squirrel management, at the present time there are no simple means of predicting total squirrel numbers during the main damage period. Moreover, there are three confounding factors which affected this study (and probably affect all such studies to a lesser or greater extent) so influencing the generality of models produced for predictive purposes. First, it is believed that the abundance and quality of foods other than tree seeds (Mollar, 1983) affected the demography of squirrels but this aspect is difficult to quantify (Ludolf, 1986). Secondly, the study area was relatively small in relation to the normal movement patterns of squirrels and resulted in a sizeable influx and efflux between the study area and surrounding habitats. Thirdly, there was an inverse relationship between trappability and food supply, as outlined above. This resulted in few captures and little information about the dynamics of the populations in several winters when seed food was available. However, the problem of winter trappability may actually be turned to the advantage of the wildlife manager. This aspect will be explored further in the last section of this paper.

TREE DAMAGE AND SQUIRREL CONTROL

Factors which influence squirrel damage and control have been reviewed by Gurnell and Pepper (1988). Three of these factors are of particular relevance here.

First, as has already been mentioned, in Britain nearly all economic damage occurs in the months of June and July. One feature worthy of note concerning the wide seasonal fluctuations in squirrel numbers is that population control has only a short-term and local effect on squirrel population dynamics (Gurnell, 1987). Therefore, control measures should only be implemented from just before and during the damage period, say from April to July. Between 1965 and 1972, only 36% of the 174 000 squirrels trapped or shot in State Forests were killed during the damage season and much control effort was wasted (Gurnell and Pepper, 1988).

Secondly, placing poison bait in damage-vulnerable habitats can be counterproductive in that squirrels are attracted to the bait in these habitats, which otherwise tend to be characterized by relatively low densities, and damage levels can increase. Where possible, squirrels should be controlled in high-density habitats, thus drawing squirrels away from damage-vulnerable habitats. Problems may arise here where habitats appear to be both damage-vulnerable and potentially high density and further research is necessary on the best procedure for the placement of hoppers in these situations.

Thirdly, more damage to trees occurs when squirrel numbers, especially juvenile numbers, are high (e.g. Kenward and Parish, 1986). Further quantitative data are required to improve our confidence in this relationship but the critical process affecting numbers during the damage period is whether winter breeding occurs and how many juveniles are recruited into the summer population. This in turn is strongly dependent on tree seed supplies, as outlined above. A logical consequence of this is that it will be unnecessary to control grey squirrels in some years when numbers are naturally low during the damage period, and indeed, although more difficult, it may be possible to vary control effort according to how much early breeding occurs.

INDEX METHOD

It has been established that few squirrels will be captured when tree seed supplies are abundant and that good seed supplies also result in an early start to breeding with production of spring litters. Moreover, adult survival is likely to be good in these circumstances. In consequence, squirrel numbers will be high during the damage period with the possibility of high damage levels. Thus, it follows that low winter trappability means high potential damage. A simple index method of assessing future relative abundance during the damage period would be to carry out a standard, pilot live-trapping study in January in potentially high-density habitats (this is not to be confused with population control). Any individuals captured will provide direct evidence of breeding status, and if few individuals are captured in traps, then samples can be supplemented by shooting or collecting road casualties. Two measures of trappability can be used in this assessment, first the number of individuals captured, and second the mean number of captures per individual (Table 12.3).

Seed crop size varies across small geographic areas (Gurnell, 1987) and for this reason the index method of future relative abundance will have to be carried out in each forest district. Based on the results obtained, the forest manager will be able to plan how much control will be required in advance of the damage season. If trappability is high, for example, the manager may consider that poison control is unnecessary that year. If trappability is very low then a full programme of control would be appropriate. Information on the severity of winter weather and tree seed supplies, in addition to the value of the tree crop (the only criterion used at present to assess whether a crop should be protected) may help the decision-making progress.

Decision risk

It is important for efficient and economic management to assess what the risk of using such an index method will be. However, it is not possible to put a monetary value on this at the moment. Biologically, the risk will depend upon:

1. The accuracy of the predictive relationships between early squirrel breeding, juvenile recruitment, population numbers and damage levels. The relationship between food availability and population processes is well founded (Gurnell, 1983). The relationship between squirrel density in potentially high-density habitats and damage levels in damage-vulnerable habitats (see Kenward and Parish, 1986), and the overlap between damage-vulnerable and potentially high-density habitats, require further research.
2. The accuracy of using trappability as an indicator of food supplies. The evidence from studies carried out in southern England suggests that trappability is a sensitive indicator of natural food abundance (Table 12.3).
3. The level of control recommended. Different levels of control can be implemented by varying the number of habitats which are poisoned within a forest and perhaps (but less likely to be implemented) by terminating the poisoning before the end of the damage period. It should be relatively easy to predict years when full control programmes (e.g. poison hoppers placed throughout all potentially high-density habitats adjacent to damage-vulnerable crops) should be implemented. It should also be possible to predict years when no control is necessary. Nevertheless, the decision risk will increase inversely with the level of control anticipated, from full down to no control (Figure 12.1). In fact, it is unclear whether it is possible to predict the levels of control necessary in different situations and studies are currently underway to test the possibility.
4. The importance of background levels of damage which may occur irrespective of squirrel density. There is little information available on this type of damage but such damage is likely to be low and unaffected by population control.

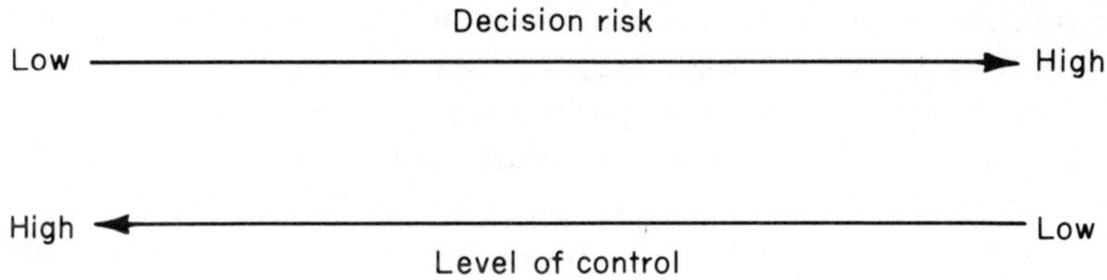

Figure 12.1 The relationship between the decision risk of recommending a particular level of squirrel control and the level of control recommended, based on an index method of assessing relative future abundance.

CONCLUSION

A summary of all grey squirrel management tactics outlined in Table 12.4. Only those factors which are directly affected by grey squirrel demography are examined in the present chapter. At the moment it can be clearly recommended that squirrels should only be controlled prior to and during the damage period and where possible only in potentially high-density habitats. The implementation of an index method for estimating

Table 12.4 Grey squirrel management tactics

1. Damage prediction assessment
   (a) Habitat damage vulnerability
   (b) Squirrel numbers during damage period, April to July
2. Time of control
   (a) Not between August and March
   (b) Not in years of low squirrel numbers (pilot sample trapping in January)
3. Place of control
   (a) In potentially high-density squirrel habitats
   (b) If possible, not in damage-vulnerable habitats
   (c) Not in relation to low yield class crops unless of high conservation value
4. Type of control
   (a) Minimize use of warfarin
   (b) Trap for small-scale control or in conservation-sensitive areas
   (c) Use squirrel-only hoppers in conservation-sensitive areas
5. Habitat management
   (a) Protect individual trees when valuable, e.g. repellents
   (b) Minimize planting of trees with large seeds or fruits, especially near damage-vulnerable crops
6. Other
   (a) Maintain a programme of monitoring and assessing damage
   (b) Research into damage economics
   (c) Research into alternative control methods to the use of poison

From Gurnell and Pepper (1988).

future relative abundance should allow a more rational and effective use of poison control by reducing or eliminating the need for control in some years. Clearly our understanding of the accuracy and implementation of the index method, and the potential risk in using it, requires careful study. The decision risk may also depend upon increasing our knowledge about the demography and spatial dynamics of squirrels in many types of habitat, such as marginal (low-density) habitats, coniferous habitats, and mixed forests. Unfortunately, studies on low-density populations are not very satisfying to the field biologist but are none the less very important. Lastly, we should not overlook the fact that damage prediction not only depends upon a knowledge of the demography and spatial dynamics of squirrels but also how these interact with the characteristics (age, species, site factors) of the trees which are damaged.

ACKNOWLEDGEMENTS

I should like to thank the Forestry Commission for permission to work at Alice Holt Forest, Surrey, and the Wildlife and Conservation Research Branch of the Forestry Commission Research Division, Alice Holt for their cooperation and assistance throughout the studies. I am also grateful to Angela Gurnell, Harry Pepper, Phil Ratcliffe, Rory Putman and a reviewer for helpful comments on the manuscript.

REFERENCES

Don, B.A.C. (1981) Spatial dynamics and individual quality in a population of the grey squirrel (*Sciurus carolinensis*). PhD thesis, University of Oxford.

Dubock, A.C. (1979) Male grey squirrel (*Sciurus carolinensis*) reproductive cycles in Britain. *J. Zool., Lond.*, **188**, 41–51.

Gosling, L.M. (1981) The dynamics and control of a feral coypu population. *Proc. Worldwide Furbearer Conf.*, **1**, 1806–25.

Gull, J. (1977) *Movement and dispersal patterns of immature grey squirrels* (Sciurus carolinensis) *in east-central Minnesota.* MS thesis, University of Minnesota.

Gurnell, J. (1981) Woodland rodents and tree seed supplies in *Proc. Wordwide Furbearer Conf.*, **1**, 1191–214.

Gurnell, J. (1983) Squirrel numbers and the abundance of tree seed supplies. *Mamm. Rev.*, **13**, 133–48.

Gurnell, J. (1987) *The Natural History of Squirrels.* Christopher Helm, London.

Gurnell, J. and Pepper, H. (1988) Perspectives on the management of red and grey squirrels in *Wildlife Management in Forests* (ed. D.C. Jardine), Institute of Chartered Foresters, Edinburgh, pp. 92–109.

Kenward, R.E. (1983) The causes of damage by red and grey squirrels. *Mamm. Rev.*, **13**, 159–66.

Kenward, R.E. (1985) Ranging behaviour and population dynamics of grey squirrels in *Behavioural Ecology: ecological consequences of adaptive behaviour* (eds R.M. Sibley and R.H. Smith), Blackwell Scientific, Oxford, pp. 319–481.

Kenward, R.E. and Parish, T. (1986) Bark stripping by grey squirrels. *J. Zool., Lond.*, **210**, 473–81.

Ludolf, C.I. (1986) *The population dynamics of rodents in a mosaic of forest habitats.* PhD thesis, University of London.

MacKinnon, K.S. (1976) Home range, feeding ecology and social behaviour of the grey squirrel (*Sciurus carolinensis* Gmelin). PhD thesis, University of Oxford.

Mollar, H. (1983) Foods and foraging behaviour of red (*Sciurus vulgaris*) and grey (*S. carolinensis*) squirrels. *Mamm. Rev.*, **13**, 81–98.

Reynolds, J.C. (1985) Autumn/winter energetics of Holarctic tree squirrels: a review. *Mamm. Rev.*, **15**, 137–50.

Rowe, J. (1980) *Grey Squirrel Control.* Forestry Commission Leaflet No. 56, HMSO, London.

Taylor, J. (1977) The frequency of grey squirrel (*Sciurus carolinensis*) communication by use of scent marking points. *J. Zool., Lond.*, **183**, 543–5.

Thirteen

Bark-stripping by grey squirrels in Britain and North America: why does the damage differ?

R.E. KENWARD

SUMMARY

Extensive surveys in British deciduous woodland showed that bark-stripping by grey squirrels tended to increase among young beech and sycamore trees when there was a high density of young animals and when the stand had been damaged in the previous year. Damage levels correlated most strongly, however, with sap volume in the vulnerable trees: their stems were never stripped extensively in an area unless there was least 0.3 ml of phloem, on average, beneath each cm^2 of bark (a 0.3 mm phloem width). A knowledge of phloem growth might therefore help to predict when damage is most likely. For example, the phloem volume is least in self-seeded saplings which grow below the main woodland canopy, and is relatively high in plantations 2–3 years after they are thinned. Most North American deciduous woodland regenerates from self-seeded understorey saplings, which probably explains why grey squirrels cause little damage there. Phloem widths averaged 0.16 mm (range of 0.08–0.26 mm) among American beech, sugar maple and red maple in 14 Canadian and US woods which were surveyed in 1986, but were 0.40 and 0.42 mm (with bark-stripping) among planted and well-thinned sugar maples in two parks. These findings suggest (1) that changes in planting practice could reduce bark-stripping in British woodland, and (2) that woodland managers in North America risk severe damage if they create maple and beech plantations.

INTRODUCTION

Grey squirrels (*Sciurus carolinensis*) can cause serious damage in Britain by stripping bark from young deciduous trees, especially beech (*Fagus sylvatica*), sycamore (*Acer pseudoplatanus*) and oak (*Quercus* spp.) which are 10–40 years old (Shorten, 1957; Rowe and Gill, 1985). The squirrels peel off and drop the outer bark from tree stems and large branches, before scraping off and eating the sap-filled phloem tissue beneath. They do this mainly in mid-summer, when phloem is especially voluminous (Kenward and Parish, unpublished data) and rich in sugars prior to lignification (Cockerham, 1930). Trees are killed outright when they are ring-barked low on the stem. Less extensive stripping can aid fatal penetration by fungi (Abbott, Bevercombe and Rayner, 1977) and insects, and stunt growth or create flaws in the harvested timber.

Squirrel management is a particular problem for foresters, because its cost must be met each year as a precaution, yet outbreaks of serious damage are sporadic: even without squirrel control, some plantations are never damaged and most others go many years with little or no bark-stripping. The damage also presents a dilemma for nature conservation. On one hand, the risk of damage and cost of its control discourages the planting of native hardwoods. On the other hand, the most effective and least-expensive prophylactic is to kill squirrels with warfarin (Rowe, 1983, Kenward *et al.*, 1988), but this may put other wildlife at risk (Wood and Phillipson, 1977).

In 1978, the Institute of Terrestrial Ecology started a project to determine the cause(s) of bark-stripping, with a view to reducing the cost of squirrel management. Ten hypotheses based on squirrel biology had been advanced to explain the damage. There was little evidence to support suggestions that stripping bark from tree stems (1) prevents tooth over-growth or (2) is an uncontrolled gnawing reflex (Davidson and Adams, 1973), that it (3) provides nest material (Brown and Yeager, 1945) or (4) water during drought (Shorten, 1954), or that the habit (5) results from a genetic mutation (Pulliainen and Salonen, 1963) or (6) from territorial boundary marking (Taylor, 1966). Support for the other hypotheses, that bark-stripping is a product of (7) agonistic gnawing

(Taylor, 1969; MacKinnon, 1976), (8) food shortage, (9) trace-nutrient deficiency (Allen, 1943) or (10) a liking for sap as a 'sweetmeat' (Mills, 1938), was sought by correlating the extent of damage with squirrel population parameters in 30 Midland woods (Kenward and Parish, 1986). Some of these sites contained small blocks of vulnerable young trees in a matrix of mature seed-bearers, and were thus areas of potential high density (PHD) for squirrels (see Chapter 12), whereas others were isolated, even-aged stands of young trees, including mixtures of deciduous and conifer species. Summer densities of squirrels, found by live-trapping, ranged from 0.3 to 9.2 per hectare. Bark-stripping was assessed on a scale of 0 (no stripping) to 5 (most trees stunted or killed). Scores of 1 (flakes $<5\ cm^2$ from a few trees) or 2 (up to $250\ cm^2$ from a few trees) were unimportant commercially, whereas 3–5 was scored by areas where more than $250\ cm^2$ was stripped from an increasing proportion of the trees.

During three study years, juvenile density was the squirrel population parameter which correlated most strongly with the level of bark-stripping. This supports the theory that bark-stripping develops during agonistic gnawing (as young squirrels establish in the population) or during exploratory feeding. There was also a relationship between the level of damage in one year and the proportion of trees gnawed the next year, even if there was little serious damage the second year. This suggested that squirrels were persistently attacking trees once they had learned the habit.

Tree quality, rather than squirrel population parameters, gave the best correlation with damage levels. Within an area, squirrels stripped the most bark from those beech and sycamore trees which gave the greatest volume of sappy phloem tissue per cm^2 of bark removed (Kenward, 1983). Single trees with less than 0.3 ml of phloem per cm^2 of bark (a 0.3 m phloem width) only had small flakes of bark removed, presumably because squirrels did not bother to strip more bark when they found little sappy tissue to eat. From 1982 onwards, the average phloem width was also recorded at each survey site. Variation in phloem width explained 74% of the variation in damage levels that year, and 58% in 1983. Phloem width and damage history data could be combined for 15 sites in 1983: together they explained 78% of the variation in 1983 damage levels (Kenward and Parish, 1986). Although its sweetness may have made the phloem an attractive food for squirrels, there was no relationship between damage levels and phloem sugar concentration. The phloem had a low energy content, and there was evidence, from squirrel weights in the survey and a feeding experiment, against any link between damage and food shortage (Kenward *et al.*, 1988).

For the management of woods vulnerable to squirrel damage, there are thus three factors which should be considered: (1) squirrel breeding and density; (2) damage history, and (3) tree quality.

Gurnell (Chapter 12) describes how squirrel breeding and density correlated with the availability of tree seeds in a 9 ha oak wood, and discusses the use of seed crops or squirrel numbers to forecast damage. In this context, it is important to note that the relationship between damage and juvenile density was non-linear (Kenward and Parish, 1986; Kenward *et al.*, 1988). When there were less than about 0.5 young squirrels per hectare, there was a tendency for sites with more squirrels to have more damage, and there were sites with no bark-stripping or with, at worst, small flakes of bark taken from a few trees. However, there was always noticeable bark-stripping when juvenile densities were 0.5 to 5 per hectare, and the variation in severity of the damage (ranging from the stripping of 5–250 cm^2 patches to the destruction of most trees) was not then related to squirrel density.

If annual damage forecasts are based on squirrel demography, it is probably only safe to relax control measures when winter seed crops fail, and provided that there was also no sign of damage the previous summer. For long-term woodland planning, beech and other trees vulnerable to stripping should not be planted near mature seed-bearing trees (Gurnell 1987 and Chapter 12), or mixed with trees which will fruit before the vulnerable age is passed. In a survey of habitat quality in 53 vulnerable woodlands, moderate to severe damage had accumulated at 31 of 44 sites which had seed-bearing trees within 200 m of the vulnerable crop, but at only 2 of 11 sites with no nearby seed-bearers (Kenward, Parish and Doyle, 1988).

Although bark-stripping may be initiated by agonistic or feeding behaviour of young squirrels, or by squirrels which have learned the habit, its severity ultimately depends on tree quality. In 186 site surveys, damage has never reached serious levels where the average phloem width was less than 0.3 mm (Kenward and Parish, 1986; Kenward, Parish and Doyle, unpublished). This raises the question of whether factors which influence phloem development can be used to forecast bark-stripping or to grow trees with reduced risk of damage. The second part of this chapter presents data on some factors which affected phloem width in British woods, and comparable data from North America which suggest why there has been little squirrel damage there.

BRITISH SITES

A number of factors may affect the width of the phloem layer in summer, which is ultimately reflected in the annual growth ring. Variation related to weather conditions and past performance may eventually be used for annual damage forecasts, but there are not yet enough data for this. Data on the effect of crowded growing conditions obtained by comparing self-seeded stands with plantations, and by comparing plantations before and after thinning.

Phloem widths were measured in 10–25 trees at 15 sites in central England on 30 June 1982. There were ten plantations, and five areas of self-seeded trees. After removing 3–5 cm^2 bark at chest height on each tree, the underlying phloem was scraped into a 1 ml syringe from which the end had been cut. In later years, a corer was used to remove 2 cm^2 of bark from each tree. This minimized sampling damage, but gave too little tissue to estimate phloem width reliably for each tree: the phloem tissue accumulated in the syringe to give an average from ten trees at each site.

The corer was used in five consecutive years to measure average phloem widths at a beech site, during a period when the crop was progressively thinned by cutting out a spruce nurse crop and removing stunted (or bark-stripped!) beech. Samples were taken in thinned and unthinned sections

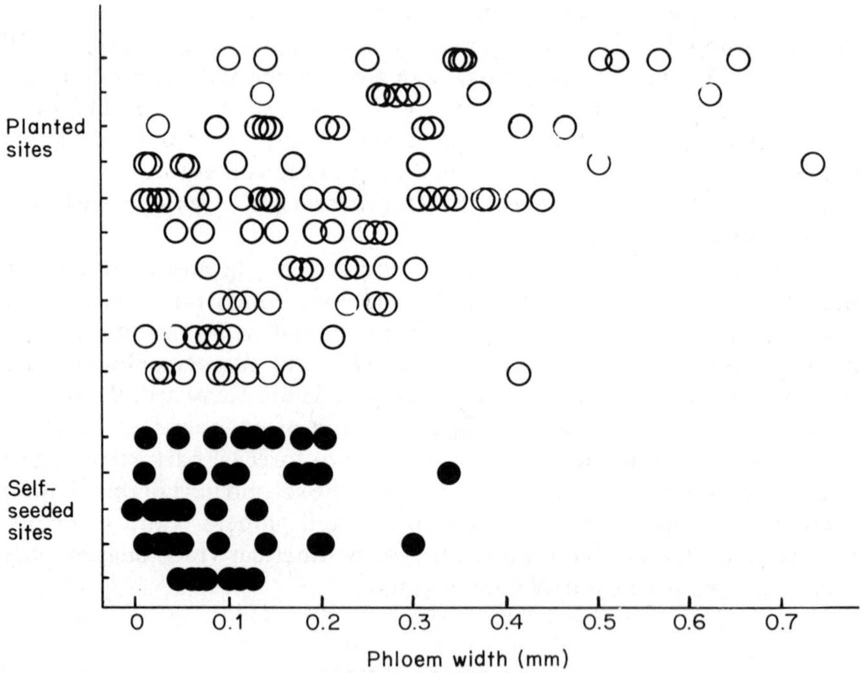

Figure 13.1 The phloem thickness on 30th June 1982 of individual trees in British plantations (○) and self-seeded woodland (●). Sites are ranked by severity of bark-stripping from top (worst) to bottom (least).

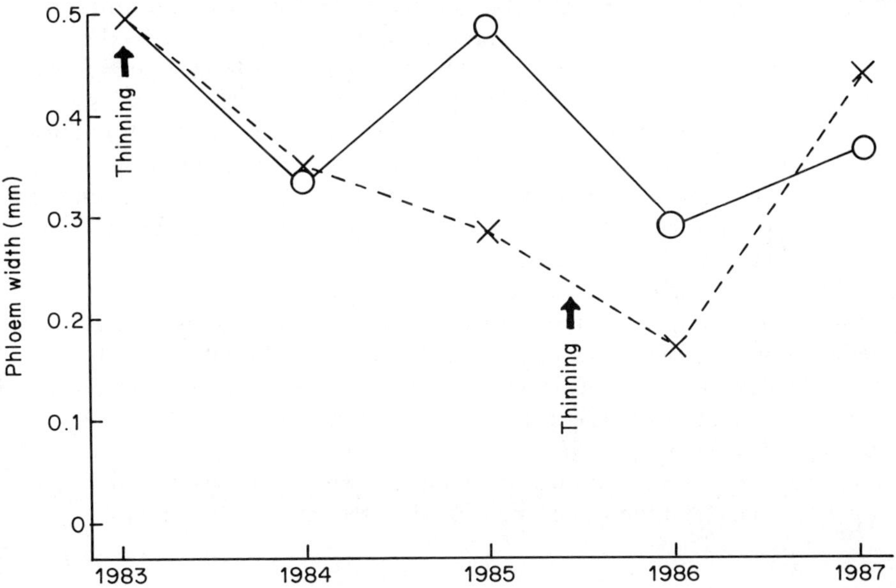

Figure 13.2 The mid-summer phloem thickness in thinned (○——○) and initially unthinned (× – – – ×) sections of a British beech plantation.

of the stand. In 1985, similar data were obtained from another beech site which had been thinned two years previously, using two cores per tree to obtain enough phloem for a measurement from each tree.

At the sites surveyed in 1982, the beech and sycamores in plantations tended to have wider phloem than the self-set trees (Figure 13.1), even though some of the plantations were in need of thinning and some self-set trees were growing strongly because the canopy had been removed. The difference between the two groups was significant if median values were compared in a Mann-Whitney test ($z = 2.327$, $p = 0.01$). Comparing individual trees, only 4% in the self-set stands had a phloem width which exceeded 0.3 mm, compared with 23% of the planted trees (Fisher test, $p = 0.0012$). The sites in Figure 13.1 are ranked by severity of bark-stripping: there was most damage where phloem widths were greatest.

Thinning the trees in plantations had a marked effect on phloem width 2–3 years later. In the site where individual trees were compared two years after thinning, the difference between samples of 15 trees in the thinned and unthinned sections was highly significant ($z = 2.616$,

$p = 0.0045$). In the site thinned in 1983 and sampled until 1987, there was no difference in phloem layers between the thinned and unthinned sections in 1984, but a quite marked difference had developed by the third year after thinning (Figure 13.2). The previously unthinned section was thinned that year, but still had less phloem than the first-thinned section in 1986, and only showed real improvement in 1987. The differences between the thinned and unthinned sections seemed mainly to be due to decreasing phloem widths in the unthinned trees, presumably due to increasing competition, whereas the thinned trees maintained their phloem layers at 0.3–0.5 mm.

The main conclusion was that crowding, whether in naturally dense self-seeded stands or in plantations ripe for thinning, was associated with reduced phloem widths. This in turn affected the vulnerability of the trees to bark-stripping. At the 1982 sites, 74% of the inter-site variation in fresh bark-stripping was explained by variation in average phloem widths (Kenward and Parish, 1986). Moreover, in a wider survey of accumulated damage at 53 sites (Kenward, Parish and Doyle, 1988), self-seeded trees were less bark-stripped than those in plantations.

AMERICAN SITES

Although grey squirrels do sometimes damage deciduous trees in North America (Britton, 1933; Brenneman, 1954; Irving and Beer, 1963), the habit is not generally considered a problem there (Gurnell, 1987). Since squirrels breed as well in North America as in Britain (Nixon and McClain, 1975; Thompson, 1978), it seemed possible that lack of damage results from differences in tree quality. For example, the trees there might have evolved antiherbivore substances to deter squirrels (1), or the phloem width might be reduced, by climatic conditions (2) or by competition for resources in American woods (3).

Sixteen sites containing American beech (*Fagus americana*), red maple (*Acer rubrum*) and sugar maple (*Acer saccharum*) were surveyed, between Patuxent in Maryland, USA and Ottawa in Canada. This represents a south–north transect through the grey squirrel range in North America. The sites were visited between 17 June and 1 July 1986, which corresponded to the timing of British surveys when seasonal advancement with latitude was taken into account. Only two sites were found where trees were well spaced, as a result of active management. The other young trees, in woods typical for the areas visited, were self-seeded saplings growing beneath the main canopy, but were generally chosen near road or field edges and thus with least competition for light and other resources.

Phloem width was measured for 10 young trees in each site, and at six sites (two for each tree species) a total of 2–3 ml of phloem tissue was taken for chromatographic analysis to detect possible aversive chemicals; the sap was also tasted at each site. As an index of squirrel presence, the trunk of each sampled tree was checked for squirrel claw marks. These show as four or five parallel scratches, each 1–3 cm long and 0.3–0.6 cm apart, running diagonally across the surface patina of the smooth-barked young trees. As many trees as possible were examined for evidence of bark-stripping, along a transect of 100–500 m through each site.

At the 14 unmanaged sites, phloem width ranged from 0.08 to 0.24 mm in American beech, from 0.08 to 0.26 mm in red maple, and from 0.07 to 0.22 mm in sugar maple (Table 13.1). None of these sites had an average phloem width greater than 0.3 mm, the threshold for extensive damage in Britain.

At least four out of ten sampled trees had been climbed by squirrels at three of four beech sites and at five of ten maple sites, but there was no sign of squirrels at three of the maple sites. At only one site, where 4 of 10 sugar maples had been climbed by squirrels, had a 10 cm^2 area been stripped freshly from the base of one tree. This tree, one of the largest saplings in woodland adjoining an Ottawa suburb, had an individual phloem width of 0.40 mm, compared with the site average of 0.09 mm. There were also signs of old damage at one beech site, where 9 of 10 trees had been used by squirrels. Hand-sized pieces had been taken from the base of two trees in previous years, but the damage could have been caused by other mammals.

Data from two managed sites near Ottawa strongly suggested that crowding, rather than climate, was responsible for the low phloem volumes in the natural woodland. The phloem width averaged 0.40 and 0.42 mm in the sugar maples at these sites. One was a small plantation of

Table 13.1 The average phloem width (mm) of young trees at sites in the North American range of the grey squirrel, sampled between 17 June and 1 July 1986

Species	Number of sites	Mean	Min.	Max.
Natural woodland				
American beech	4	0.16	0.08	0.24
Red maple	5	0.20	0.08	0.26
Sugar maple	5	0.12	0.07	0.22
Managed woodland				
Sugar maple	2	0.41	0.40	0.42

trees in a park, far from other woodland and with no sign of squirrels. The other site, again at the edge of a park, contained self-set saplings where mature trees had been removed. Squirrels had freshly removed a few flakes of bark, and some trees had been damaged quite severely 2–3 years before the survey.

There was no bitter or acrid taste to any of the saps sampled. Compared with phloem from British beech and sycamore, there was no unusual concentration of chemicals in the parts of the chromatogram spectrum usually associated with aversive compounds (Kenward and Wilson, unpublished). However, after several sites had been visited, it became clear that the outer bark of the sampled red and sugar maples was generally thicker than that of sycamores and Norway maples (*Acer platanoides*) in Britain, and this may have deterred the squirrels.

DISCUSSION

The results strongly suggest that growing conditions in naturally regenerating woodland are the main hindrance to bark-stripping in North American woods, by producing thin phloem layers. The removal of bark flakes was recorded at three of the 16 sites visited, but the low phloem volume would not have encouraged squirrels to strip larger areas. Squirrels probably remove bark flakes quite frequently in North American woods, without this necessarily being noticed by foresters (the scar is small, and usually heals over with 2 years). However, there may also be less initiation of damage than in Britain, partly because prolonged snowcover or natural predators may reduce the success of spring breeding in some areas, and partly because few squirrels will be rewarded with enough phloem to learn the habit.

Although American trees appeared not to contain aversive chemicals, rapid thickening of the bark probably provided an alternative defence against bark-stripping. There may be scope for research to develop thick-barked strains of vulnerable British species, but bark-thickening does not provide complete protection. Young oak have relatively thick bark and are damaged less by squirrels than are sycamore and beech (Shorten, 1957), but the bark is sometimes peeled where it is thin, high up the young trees. Loss of the crown, through ring-barking in the canopy, can be a problem in this species. In the long term, there may still be scope for the insertion of genes for antifeedants into beech or other vulnerable native trees.

More research is needed before the findings on tree quality can be fully exploited in forestry, but they certainly offer new possibilities for reducing squirrel damage. In terms of damage forecasting, risk is likely to be high 2–3 years after a plantation is thinned, and in areas where tree growth is rapid. Could we refine our forecasts by using weather conditions to predict when phloem volume will be too low for squirrel damage, at least in some sites? Could we manage tree growth so that phloem layers remain thin until the trees are too old and the bark too thick for serious damage? Would planting new trees under mature canopy help in this respect? Is it practical to reduce squirrel damage in nature reserves, where rapid tree growth may be unimportant, by deliberately crowding the beech seedlings?

Whatever the eventual answers to these questions, the role of tree quality in bark-stripping needs more attention by foresters in both Britain and North America. In Britain, for instance, there is an increasing tendency to plant trees with a wide spacing. This may reduce planting and thinning costs, but it will also maximize the risk of bark-stripping. In North America, the development of commercial broadleaf plantations might well produce as much bark-stripping as in Britain.

CONCLUSIONS

In Britain, serious bark-stripping by grey squirrels is rare unless average phloem width in vulnerable plantations exceeds 0.3 mm. In self-seeded British woods, only two trees out of 50 had phloem widths greater than this, compared with 27 of 115 trees in plantations. Phloem width was also greater in plantations 2–3 years after thinning than in nearby unthinned trees. In 14 natural woodland sites in North America, average phloem widths were all below 0.3 mm, probably because of competitive growing conditions in the self-seeding woods: well-spaced sugar maples at two managed sites had phloem layers which averaged 0.41 mm, and one of the sites had been damaged by squirrels. It is suggested that squirrel damage will be increased by the recent practice of planting trees well-spaced to reduce management, and by any move towards plantation forestry in North America.

ACKNOWLEDGEMENTS

For help with different aspects of this work, I thank F.I.B. Doyle, M.R. Fuller, J.T. Kenward, G. Kenward, J.A. Mosher, T. Parish, J. Ruos, M. Wilson and the Forestry Commission. This chapter was improved by comments from M.G. Morris, R.J. Putman and an anonymous referee.

REFERENCES

Abbott, R.J., Bevercombe, G.P. and Rayner, D.M. (1977) Sooty bark disease of sycamore and the grey squirrel. *Trans. Br. Mycol. Soc.*, **69**, 507–8.

Allen, D.L. (1943) Michigan fox squirrel management. *Mich. Dept. Cons. Game Divn Publ.*, **100**, 1–404.

Brenneman, W.S. (1954) Tree damage by squirrels silviculturally significant? *J. For.*, **52**, 604.

Britton, W.E. (1933) Injury to trees by squirrels. *Nat. Shade Tree Conf. Proc.*, **9**, 85–91.

Brown, L.G. and Yeager, L.E. (1945) Fox squirrels and gray squirrels in Illinois. *Bull. Ill. Nat. Hist. Surv.*, **23**, 445–536.

Cockerham, G. (1930) Some observations on cambial activity and seasonal starch content in sycamore (*A. pseudoplatanus*). *Proc. Leeds Phil. Lit. Soc.*, **2**, 64–80.

Davidson, A.M. and Adams, W. (1973) The grey squirrel and tree damage. *Q.J. For.* **67**, 237–61.

Gurnell, J. (1987) *The Natural History of Squirrels*. Christopher Helm, London.

Irving, F.D. and Beer, J.R. (1963) A six year study of sugar maple bark stripping by gray squirrels in a Minnesota oak maple stand. *J. For.*, **61**, 508–11.

Kenward, R.E. (1983) The causes of damage by red and grey squirrels. *Mamm. Rev.*, **13**, 159–66.

Kenward, R.E. and Parish, T. (1986) Bark-stripping by grey squirrels (*Sciurus carolinensis*). *J. Zool. Lond.* (A) **210**, 473–81.

Kenward, R.E., Parish, T. and Doyle, F.I.B. (1988) Grey squirrel bark-stripping. II. Management of woodland habitats. *Q.J. For.*, **42**, 87–94.

Kenward, R.E., Parish, T., Holm, J. and Harris, E.H.M. (1988) Grey squirrel bark-stripping. I. The roles of tree quality, squirrel learning and food abundance. *Q.J. For.*, **42**, 9–20.

MacKinnon, K.S. (1976) Home range, feeding ecology and social behaviour of the grey squirrel (*Sciurus carolinensis*. Gmelin). PhD thesis, Oxford University.

Mills, E.M. (1938) Tree injury by squirrels. *Bull. Mass. Agr. Exp. Stn.*, **353**, 79–80.

Nixon, C.M. and McClain, M.W. (1975) Breeding seasons and fecundity of gray squirrels in Ohio. *J. Wildl. Mgmt*, **39**, 426–38.

Pulliainen, E. and Salonen, K. (1963) On ring-barking of pine by the squirrel (*Sciurus vulgaris*) in Finland. *Ann. Acad. Sci. Fenn.*, **72**, 1–29.

Rowe, J.J. (1983). Squirrel management. *Mamm. Rev.*, **13**, 173–82.

Rowe, J.J. and Gill, R.M.A. (1985) The susceptibility of tree species to bark-stripping damage by grey squirrels (*Sciurus carolinensis*) in England and Wales. *Q.J. For.*, **79**, 183–90.

Shorten, M. (1954) *Squirrels*. Collins, London.

Shorten, M. (1957) Damage caused by squirrels in Forestry Commission areas 1954–6. *Forestry*, **30**, 151–72.

Taylor, J.C. (1966) Home range and agonistic behaviour in the grey squirrel. *Symp. Zool. Soc. Lond.*, **18**, 229–35.

Taylor, J.C. (1969) Social structure and behaviour in a grey squirrel population. PhD thesis. University of London.

Thompson, D.C. (1978) Regulation of a northern grey squirrel (*Sciurus carolinensis*) population. *Ecology*, **59**, 708–15.

Wood, D.A. and Phillipson, J. (1977) The utilisation of poison hoppers designed for grey squirrel (*Sciurus carolinensis* Gmelin) control. *Biol. Conserv.*, **11**, 119–27.

Fourteen

Demographic consequences of differences in the ranging behaviour of male and female coypus

L.M. GOSLING AND S.J. BAKER

SUMMARY

Coypus have a polygynous mating system: females occur singly or in clusters of related individuals in patches of wetland habitat and males attempt to monopolize a number of females. Male ranges are thus larger than those of females and radio-tracking shows that they move at a faster rate. Males also disperse to new ranges further and more frequently than females. Collectively, these behaviours should make males more likely to encounter traps and a comparison of capture rates (numbers trapped in relation to estimated number available) showed that males were caught at a rate 22% higher than females. Consequently, male survivorship was relatively poor, particularly at high trapping intensity, and the adult sex-ratio became progressively biased towards females as trapping intensity increased. Cold weather also directly reduced male survival although this effect was smaller than that of trapping. When the coypu population was reduced to very low density in an intense trapping campaign, the rarity of males may have limited population fecundity. This density-dependent response appears to have increased the chance that the trapping campaign will succeed in eradicating coypus from Britain.

INTRODUCTION

A population of feral coypus (*Myocastor coypus* Molina) has inhabited the wetlands of East Anglia since the 1930s when a few escaped, or were released, from fur farms (Laurie, 1946; Norris, 1967a; Gosling, Watt and Baker, 1981). Since the 1950s when they became numerous (possibly reaching 200 000: Norris, 1967a) they have been regarded as pests because of their damage to crops, drainage systems (Cotton, 1963; Norris, 1967a) and to native plant communities (Boorman and Fuller, 1981; Ellis, 1963). They have been systematically controlled by trapping since 1962 (Norris, 1967a; Gosling, Watt and Baker, 1981). An attempt to eradicate the feral population started in 1981 and estimated numbers of adult females declined from about 3000 in April 1981 to about 20 in April 1986 (Gosling and Baker, 1987). The only technique used since 1962 is cage-trapping (Norris, 1967b); traps are set on runs which coypus use as they move about their ranges, or on baited rafts which coypus find as they swim along water courses. The number of trappers has varied between a minimum of five in the late 1960s and a maximum of 25 during the eradication campaign in the 1980s. Trapping effort during the eradication campaign was over 200 000 trap-nights per year (Gosling and Baker, 1987).

The first aim of this chapter is to describe differences in the behaviour of male and female coypus and to consider the way in which these differences could affect the chances of capture during the trapping campaign. Like most mammals, coypus have a polygynous mating system (Warkentin, 1968; Gosling and Baker, 1989). In such systems males may be more conspicuous, and thus more easily caught, because of their greater activity during attempts to find and monopolize a number of mates. Next, we explore the effect of differences in catchability on the survival of males and females and the effect on the adult sex-ratio. Males may also be more vulnerable than females to cold weather and we carry out a preliminary analysis to assess the relative importance of this factor on the adult sex-ratio.

Average fecundity in a population of coypus is unlikely to be affected by the proportion of males because each male can potentially fertilize a large number of females. However, this generalization may not apply when females are very rare and when they are scattered over large areas. Under these circumstances, the potential for polygyny could decline to very low levels and the abundance of males could limit population fecundity. In this paper we discuss whether such events could have occurred when the coypu population was reduced to low levels by trapping and consider the implications for other attempts to remove populations of introduced mammals.

MATERIALS AND METHODS

Home ranges were estimated for coypus kept under natural conditions in an 18 ha enclosure which contains Calthorpe Broad in Norfolk (Gosling and Baker, 1980). 150 traps were used, mostly positioned on the banks of ditches and the broad. The coypus were trapped once a month between 1970 and 1979 and twice a month from 1979 to 1987. Individuals were marked by clips in the webs of the hind feet and by ear-tags. The ranges of adults were defined as the areas covered when each individual was over six months old. Ranges were also determined in a number of studies carried out on unenclosed wild coypus in a variety of habitats in Norfolk. Most of this work involved trapping, as at Calthorpe, but some ranges were also determined using radio-tracking (see below). Ranges were calculated from the area within a convex polygon joining the outermost observations of each individual.

Rates of movement were determined in a four-year study (1982–1986) of social behaviour using radio-telemetry. The study area was a 480 ha site consisting mainly of wetland pasture in the River Yare valley in Norfolk. All coypu caught during four days of trapping each month were fitted with a collar holding a radio transmitter. Each animal was tracked for three nights each month using two Land Rovers, each equipped with a telescopic mast, an Adcock antenna and a repeater compass system (Cederlund and Lemnell, 1980). The position of the subject, and any other animal that could be located from the same tracking positions, was determined at 30 minute intervals by triangulation from the two receiving stations and recorded to an accuracy of 100 m as a six figure grid reference.

Information about dispersal away from a stable range was obtained from 449 coypus which were individually marked, and recaptured, in various locations throughout Norfolk between 1972 and 1987. Animals were marked for various reasons, for instance for radio-telemetry and during a study of immigration into an intensively trapped area (Gosling, Baker and Clark, 1988). They were often killed during control trapping by trappers of the Coypu Control Organization. The values used here are the distances between the first and last capture position when these were over 30 days apart.

Age structures were obtained from animals supplied by the Coypu Control Organization for post-mortem examination. Aging was from eye lens weight (Gosling, Huson and Addison, 1980). Since 1970, the population has been monitored by reconstructing the numbers of adults, using the numbers and ages of the animals killed each month (Gosling, Watt and Baker, 1981; Gosling and Baker, 1987); the population analyses in this paper refer to events between 1970 and 1986. Details of the number of coypus killed each month and the trapping effort expended were supplied by the Coypu Control Organization.

RANGING BEHAVIOUR

Female coypus occupy patches of fen, or related habitats either singly or in clusters. Clusters are formed when daughters remain near to their mothers when they become adult in partly overlapping ranges; the resulting aggregation is thus a kin-group. The ranges of single large males overlapped groups of females or, at low density, the ranges of a number of isolated females. These observations are consistent with the possibility that males attempt to monopolize and mate with a number of females, as in other polygynous mammals (Gosling and Baker, 1989).

There was great variation in range size measured under different circumstances (Table 14.1). Large ranges were most common on grazing marsh sites (where the habitat suitable for coypus is confined to the ditches but the range determined by the minimum polygon method includes the fields in between) and at low population density. Although ranges at Calthorpe were smaller on average than those elsewhere, ranges

Table 14.1 Range areas of male and female coypus. The areas are those within a concave polygon joining the outermost points where each animal was observed. Observations were from trapping and radio-tracking. Ranges with less than ten observations are excluded. The ranges of coypus in France are from Doncaster and Micol (1989)

Sample	Mean range areas (± sd; ha)		
	Males *n*	Females *n*	*p*
1. Calthorpe enclosure:	8.4 ± 3.0 (55)	5.5 ± 3.2 (68)	<0.001
2. Strumpshaw (unenclosed):	6.8 ± 5.5 (8)	3.0 ± 1.8 (7)	0.118
3. Grazing marsh (unenclosed):	93.9 ± 69.5 (14)	46.3 ± 30.0 (16)	0.018
4. France (unenclosed):	5.7 ± 4.0 (8)	2.5 ± 1.8 (11)	0.019

The means in samples 1–3 are compared using a Mann–Whitney U test and those in sample 4 using Wilcoxon's two-sample test (Doncaster and Micol, 1988).

in comparable habitats (mixed aquatic, fen and carr habitats) were of similar sizes. In France, Doncaster and Micol (1989) found an overall mean range size for wild adults of 3.8 ± 3.3 ha which is comparable with an overall range of 6.8 ± 3.4 ha for adults at Calthorpe.

Males proved to have significantly larger ranges than females in all of the samples (Table 14.1).

Although males have larger ranges than females they would not encounter traps more frequently unless they moved throughout their ranges at a rate which was at least the same as that of females. We tested the rate at which males and females moved by radio-tracking (see Materials and methods). The position of each animal was determined at 30 min intervals and the distance moved between successive positions was estimated later. On average, males moved further than females (Figure 14.1).

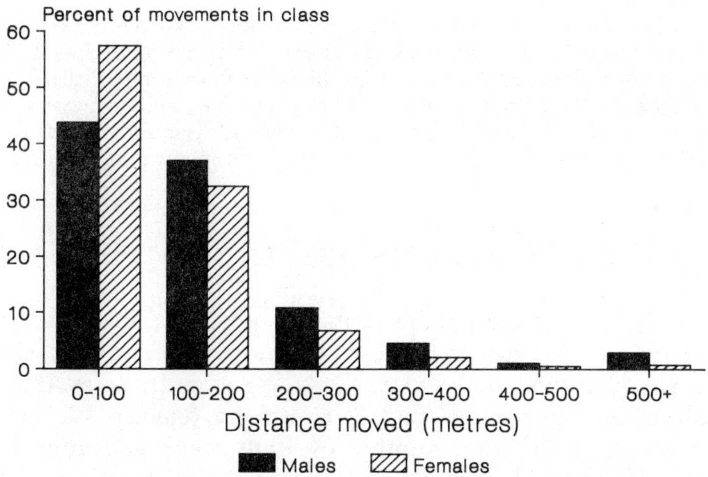

Figure 14.1 Distances moved by radio-tracked adult male and female coypus at 30 minute intervals (see Materials and methods section for details).

We compared the frequency and distance of dispersal away from stable ranges (probably natal ranges in some cases) from the distances between the positions where animals were marked and recaptured (see Materials and methods). Most animals stayed near the point where they were caught but the information on animals which dispersed (Figure 14.2) suggests that males disperse more frequently than females and that they move larger distances. Similar results have been obtained in many studies of small mammals (e.g. Howard, 1960; Smith, 1965).

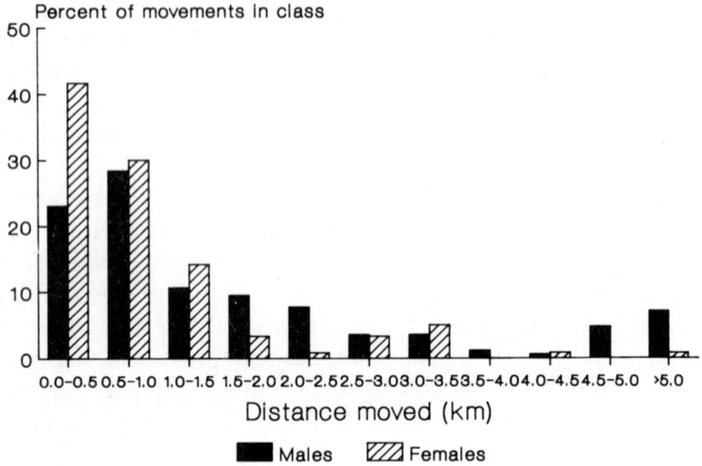

Figure 14.2 The distances moved by adult male and female coypus between marking and eventual recapture. Time intervals of less than 30 days were excluded. The data show that coypus remained in the vicinity of the first capture: in these clustered observations, males range more widely, reflecting their larger home ranges (see Table 14.1). Movements beyond the cluster probably indicate dispersal movements to new ranges and males undertook such movements more commonly than females.

CAPTURE RATES AND LONGEVITY

Males thus have larger ranges than females, they move about their ranges more quickly and they disperse further and more often. Collectively, these behaviours could make males more likely to encounter traps. If this is so then males should be caught at a faster rate than females. We tested this using estimates of the total number of adult males and adult females trapped each month in relation to estimates of the number alive each month from population reconstruction (Gosling, Watt and Baker, 1981; also see Materials and methods). On average males were trapped at a rate 22% higher than females (males: 0.16 ± 0.07, females: 0.13 ± 0.06, paired $t = 8.52$, 16 df, $p < 0.001$). The capture rate increased significantly with trapping intensity (Figure 14.3). There was also a suggestion that males were caught at a relatively higher rate at high trapping intensity but the male and female slopes in Figure 14.3 were not significantly different.

If adult males are killed at a higher rate their longevity will be less than that of females. The extent of this effect was estimated using a survivorship analysis for a sample of 14,221 coypus which had reached six months of age. The sample was divided into animals from years with greater or less than median trapping intensity and the data in each block

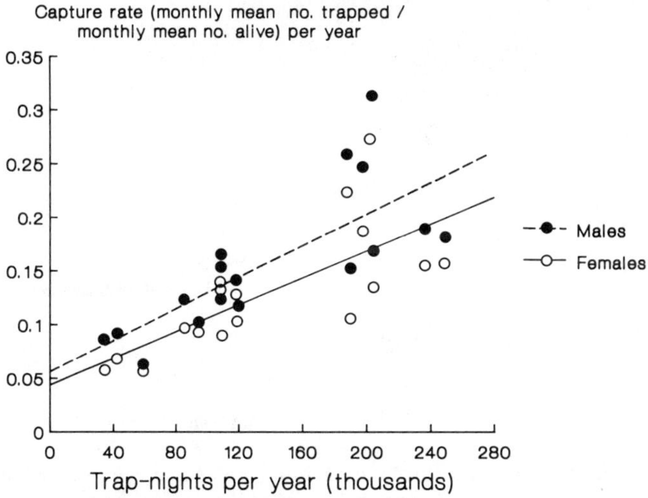

Figure 14.3 The capture rate of adult male and adult female coypus in relation to variation in trapping intensity. Annual capture rates were obtained from the numbers of each sex trapped each month divided by the number alive (from population reconstruction). These values were averaged for each of the 17 years from 1970 to 1986. Trapping intensity is the total number of trap nights each year.

were expressed as differences in survival between the sexes in relation to age. A comparison of the curves obtained (Figure 14.4) suggests that increased trapping intensity has a greater effect on male than on female survival.

ADULT SEX-RATIOS

On average there are more male neonates born per litter than females (males: 2.89 ± 1.46, females: 2.53 ± 1.52; $n = 741$ litters, $t = 4.39$, $p < 0.001$; Gosling, 1986). At six months of age more males are being trapped than females (Figure 14.5) and this difference may be partly due to the sex-ratio at birth and partly because males are already more vulnerable to trapping, particularly when they disperse as young adults. After about 15 months females are caught more frequently than males, probably because the male sector of the population has been depleted by trapping. Another factor which affects the adult sex-ratio is that males become sexually mature later than females (males 4–8 months; females 3–8 months; Gosling and Baker, 1988) and this tends to reduce the proportion of males amongst adults.

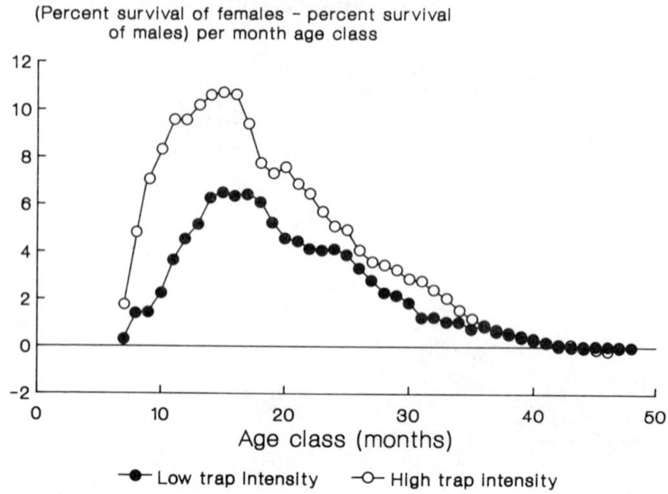

Figure 14.4 Differences in the survival of adult males and females at high and low trapping intensity. The sample consisted of all aged males and females which were trapped and examined between 1970 and 1986 (n = 7535 males and 6686 females). These were divided into two groups consisting of years above and below median trapping intensity (total trap-nights per year). Within each group animals were grouped in three month classes and survival from six months of age was calculated in the usual way for each sex. The curves shown here are the differences between the male and female curves (expressed in percentage survival terms) for the high and low trapping intensity groups.

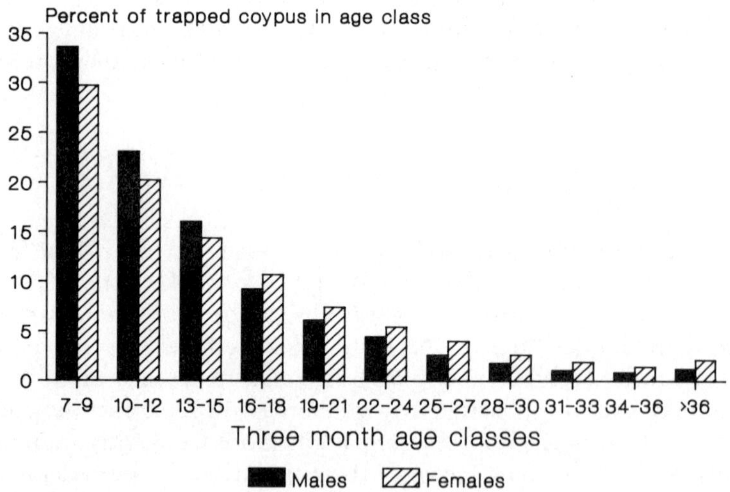

Figure 14.5 The number of adult males and females trapped in three-month age classes. To avoid the affect of different ages of recruitment in the sexes all animals over six months old are included. The sample consists of all animals over six months which were examined between 1970 and 1986; n = 7535 males and 6686 females.

Table 14.2 Multiple regression analysis of the effect of trapping intensity and winter severity on the proportion of males in the adult sector of the population. The proportion of males was obtained from a monthly reconstruction of the population (see text, and Gosling, Watt and Baker, 1981) averaged for each of the 17 years from 1970 to 1986. Winter severity was expressed using an index which weights and accumulates runs of freezing days each winter (see Gosling, 1981); trapping intensity is the total number of trap-nights each year

Independent variable	Beta	*t* value	*p*
Trapping intensity	−0.61	−4.05	0.001
Winter severity	−0.41	−2.78	0.015

Coefficient of determination: 0.70; $F = 19.6, p = 0.0001$.

The effect of trapping on the adult sex-ratio was explored by regressing trapping intensity against the proportion of adult males in the reconstructed population in each year from 1970 and 1986. There was also some indication from time series analysis that adult males may sometimes die from the direct effects of very cold winters and winter severity was also entered as an independent variable in a multiple regression analysis. Both trapping intensity and winter severity proved to have significant effects on the sex-ratio although the contribution of weather was small compared to that of trapping (Table 14.2). At low trapping intensity the sex-ratio is close to unity but as trapping intensity increases there is a linear decline in the proportion of males (Figure 14.6); the trend in the bivariate relationship with winter severity is less clear, presumably due to the unrelated, larger effects of variation in trapping. Together, trapping and cold weather explained 70% of the variation in adult sex-ratio.

DISCUSSION

In polygynous mating systems, males may be generally more conspicuous as they attempt to find and monopolize a number of mates. Female reproductive success is limited by the number of young that can be produced and raised, and since these efforts will be primarily food-limited, their behaviour should be mainly directed towards finding and protecting a food supply for reproduction. Searching for mates and intrasexual competition may account for most of the differences in behaviour described here: males have larger ranges, perhaps so that they can get access to a number of overlapping female ranges and they move

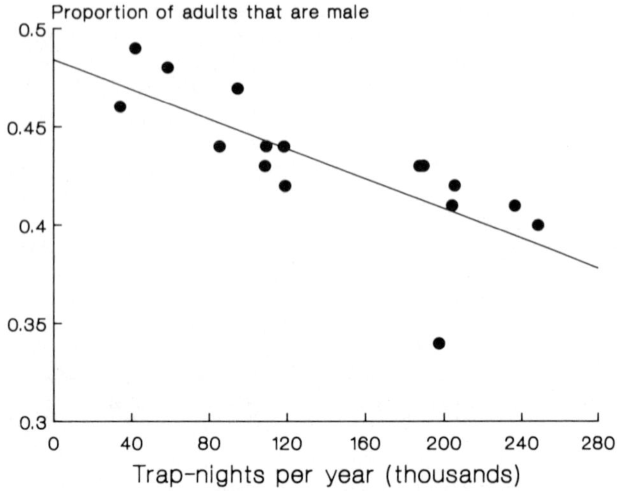

Figure 14.6 The effect of trapping intensity on the proportion of males in the adult sector of the population. The variables are described in the legend to Table 14.2. When winter severity was held constant in a partial correlation procedure, the coefficient for the relationship between trapping intensity and the proportion of males was −0.73 (p = 0.001).

around their range more quickly to visit these females (either to monitor their reproductive success or to prevent cuckoldry by other males). Dispersal to a new range may also be primarily concerned with competition for mates (e.g. Greenwood, 1980; Dobson, 1982) although inbreeding avoidance is sometimes important (Pusey, 1987).

Because of these differences in behaviour, males should be more likely to encounter traps and they should thus be caught at a higher rate than females. The results presented here suggest that this is so. Similar male-biased results have been obtained in a number of small mammal trapping studies (for example *Peromyscus polionotus* and *Mus musculus*; Smith, 1968) and in some of these there is a tendency for the proportion of males caught to increase in the summer when differences in male and female behaviour might be greatest (e.g. *Apodemus sylvaticus* and *Clethrionomys glareolus*; Gurnell, 1982; Southern, 1973).

At birth there is commonly an excess of males in polygynous mammals (Clutton-Brock and Albon, 1982) but, because males are more vulnerable to most mortality factors (reviewed by Trivers, 1985), female-biased ratios are the rule amongst adults. Sometimes the differential mortality of males has been explicitly linked to male intrasexual competition (e.g. *Microtus montanus* with sex ratios of up to 1 : 1.38; Jannett, 1981).

However, other factors are sometimes important: cold weather was shown to have a significant negative effect on the proportion of male coypus in the population although the effect was smaller than that of trapping. An analysis of the causes of this effect is outside the scope of this chapter but it is important to point out that there are a number of possibilities. These include selective mortality of male juveniles (which would then affect recruitment to the adult sector of the population), direct differential mortality on adult males, and changes in male behaviour due to cold weather which would make males more likely to be trapped; these possibilities will be considered elsewhere.

The main consequences of a higher rate of capture of males compared with that of females are to reduce male longevity and, as trapping intensity increases, to progressively reduce the proportion of adult males in the population. But, are there any circumstances when the rarity of males limits population fecundity? In general, this seems unlikely: the reproductive success of female mammals is limited by the number of young that can be produced during its life: each offspring (or litter) requires considerable investment, particularly during pregnancy and lactation, and the female can only produce one offspring (or litter) at a time. Males are limited only by the number of matings that they can achieve and so a successful male can potentially achieve relatively higher reproductive success than a successful female. For this reason, it is generally assumed that the availability of males does not limit population fecundity: even if only a small number of males are present in an area they should usually be able to mate with most females.

However, the generalization that males never limit population fecundity may not apply when a population is declining to very low levels, as in the case of coypus after 1985 (Gosling and Baker, 1987). When females are rare and scattered over a large area the potential for polygyny will be low; some males may have access to only one or a very few females and some may be completely isolated from potential mates. Under these circumstances it would be necessary to have an adult sex-ratio of at least unity for population fecundity to be maintained at previous levels. When males are less common than females because they are killed at a higher rate by trapping and, to a lesser extent, by cold weather, they could thus limit population fecundity. This could have occurred by early 1985 when, after six years of the coypu eradication programme, the adult sector of the population had been reduced to less than 400 animals over an area of some 12 020 km^2. At this time there was an increase in the number of females that failed to conceive (Gosling and Baker, 1987) and these events may be linked.

The main problem anticipated when deciding whether or not to attempt to eradicate coypus from Britain was that coypus might become difficult to catch when reduced to very low density. There was concern that this factor, perhaps compounded by unpredictable changes in behaviour at low density, might prevent further reduction in population size (MAFF, 1978). In the event, the density-dependent reduction in fecundity which has been discussed above, appears to have enhanced the chances of success of the eradication campaign. This result may give encouragement to those considering attempts to remove introduced mammals where these are causing environmental damage.

REFERENCES

Boorman, L.A. and Fuller, R.M. (1981) The changing status of reedswamp in the Norfolk Broads. *J. Appl. Ecol.*, **18**, 241–69.

Cederlund and Lemnell (1980) A simplified technique for mobile radio tracking in *A Handbook on Biotelemetry and Radio Tracking*, (eds C.J. Amlaner and D.W. Macdonald), Pergamon Press, Oxford, pp. 453–6.

Clutton-Brock, T.H. and Albon, S.D. (1982) Parental investment in male and female mammals in *Current Problems in Sociobiology* (eds King's College Sociobiology Group), Cambridge University Press, Cambridge, pp. 223–47.

Cotton, K.E. (1963) The Coypu. *River Boards Assoc. Ybk*, **11**, 31–9.

Dobson, F.S. (1982) Competition for mates and predominant juvenile male dispersal in mammals. *Anim. Behav.*, **30**, 1183–92.

Doncaster, C.P. and Micol, T. (1989) Annual cycle of a coypu population: male and female strategies, (in press, *J. Zool.*, London).

Ellis, A.E. (1963) Some effects of selective feeding by the coypu (*Myocastor coypus*) on the vegetation of Broadland. *Trans. Norf. Nor. Nat. Soc.*, **20**, 32–5.

Gosling, L.M. (1981) Climatic determinants of spring littering by feral coypus, *Myocastor coypus. J. Zool. Lond.*, **195**, 281–8.

Gosling, L.M. (1986) Selective abortion of entire litters in the coypu; adaptive control of offspring production in relation to quality and sex. *Am. Nat.*, **127**, 772–95.

Gosling, L.M. and Baker, S.J. (1980) Acidity fluctuations at a broadland site in Norfolk. *J. Appl. Ecol.*, **17**, 479–90.

Gosling, L.M. and Baker, S.J. (1987) Planning and monitoring an attempt to eradicate coypus from Britain. *Symp. Zool. Soc. Lond.*, **58**, 99–113.

Gosling, L.M. and Baker, S.J. (1989) Coypu in *The Handbook of British Mammals*, 3rd edn, (eds G.B. Corbet and S. Harris) Blackwell Scientific Publications, Oxford, in press.

Gosling, L.M., Baker, S.J. and Clarke, C.N. (1988) An attempt to remove coypus (*Myocaster coypus*) from a wetland habitat in East Anglia. *J. Appl. Ecol.*, **25**, 49–62.

Gosling, L.M., Huson, L.W. and Addison, G.A. (1980) Age estimation of coypus (*Myocastor coypus*) from eye lens weight. *J. Appl. Ecol.*, **17**, 641–7.

Gosling, L.M., Watt, A.W. and Baker, S.J. (1981) Continuous retrospective census of the East Anglian coypu population between 1970 and 1979. *J. Anim. Ecol.*, **50**, 885–901.

Greenwood, P.J. (1980) Mating systems, philopatry and dispersal in birds and mammals. *Anim. Behav.*, **28**, 1140–62.

Gurnell, J. (1982) Trap responses in woodland rodents. *Acta Theriol.*, **27**, 123–37.

Howard, W.E. (1960) Innate and environmental dispersal of individual vertebrates. *Am. Midland Nat.*, **63**, 152–61.

Jannett, F.J. (1981) Sex ratios in high-density populations of the montane vole, *Microtus montanus*, and the behaviour of territorial males. *Behav. Ecol. Sociobiol.*, **8**, 297–307.

Laurie, E.M.O. (1946) The coypu (*Myocastor coypus*) in Great Britain. *J. Anim. Ecol.*, **15**, 22–34.

MAFF (1978) *Report of the Coypu Strategy Group*. Ministry of Agriculture, Fisheries and Food, London.

Norris, J.D. (1967a) The control of coypus (*Myocastor coypus* Molina) by cage trapping. *J. Appl. Ecol.*, **4**, 167–89.

Norris, J.D. (1967b) A campaign against feral coypus (*Myocastor coypus* Molina) in Great Britain. *J. Appl. Ecol.*, **4**, 191–9.

Pusey, A.E. (1987) Sex-biased dispersal and inbreeding avoidance in birds and mammals. *Trends. Ecol. Evol.*, **2**, 295–9.

Smith, M.H. (1965) Dispersal capacity of the dusky-footed woodrat, *Neotoma fuscipes*. *Am. Midland Nat.*, **74**, 457–63.

Smith, M.H. (1968) A comparison of different methods of capturing and estimating numbers of mice. *J. Mamm.*, **49**, 455–62.

Southern, H.N. (1973) A yardstick for measuring populations of small mammals. *Mamm. Rev.*, **3**, 1–10.

Trivers, R.L. (1985) *Social Evolution*. Benjamin/Cummings Publishing, Menlo Park, California.

Warkentin, M.J. (1968) Observations on the behavior and ecology of the nutria in Louisiana. *Tulane Stud. Zool.*, **15**, 10–15.

Fifteen

Rabbits as pests of winter wheat

M.J. CRAWLEY

SUMMARY

Experiments were carried out over a three-year period on the effects of the timing and duration of grazing on winter wheat, using natural and confined populations of rabbits. Experimental manipulation of grazing intensity was not attempted.

Yield declined as an approximately linear function of the duration of exposure to rabbit grazing. There was no evidence of yield compensation for early defoliation, despite vigorous regrowth by the wheat after fencing.

Exposure for a fixed period of 60 days was most damaging when it occured at the beginning of the growth period (November and December). Yield losses were minimal (but statistically significant) when the crop was exposed for 60 days just prior to harvest. Yield losses from 30-day exposure were roughly equal for each of the winter months from November to March.

Grazing caused reductions in ear density, ear weight, seed number and individual seed weight. In order of importance, yield losses resulting from 30-day exposure in November 1985 were: (1) reduced numbers of ears per unit area due to shoot mortality (38.5% loss); (2) reduced ear size (18.4% loss); which is due to (3) reduced numbers of grains per ear (7% loss); and (4) reduced individual mean grain weights (22.3% loss). Continuously grazed plots produced a few tiny ears, many of which contained only one seed.

Indirect effects of rabbit grazing include increased weediness and increased damage from cereal aphids. Split plot trials with extra herbicide applied during the rapid growth phase, and aphicide applied at the time of ear formation, produced significant interaction effects with defoliation. Herbicide increased yields only on the grazed plots, while aphicide increased yields only on plots that were both grazed and treated with herbicide.

Despite the lack of compensation for early defoliation, the wheat plants exhibited extraordinary resilience to repeated defoliation by rabbits. Even on the continuously grazed plots, some plants were able to ripen relatively large ears, containing heavy grains. There is considerable scope for experimentation aimed at devising cultural methods to maximize the proportion of the grazed crop that regrows in this way.

INTRODUCTION

By the early 1980s it was clear that rabbits had returned to their pre-myxomatosis numbers in some areas and that, over the country as a whole, rabbits were the principal pests of winter wheat, accounting for losses of more than £100 million per year (Mills, 1986). The present study was carried out over three years, and set out to determine the effects of rabbit grazing on the growth, survival and yield of winter wheat (Crawley and Brown, 1988, a,b). The aim was to discover the effects of the timing, duration and intensity of defoliation on the performance of winter wheat variety 'Hustler'. In attempting to design an experimental protocol that would allow the level of grazing intensity (the average rate of rabbit feeding per plant) to be varied, it became clear that this would require large, replicated enclosures, where rabbit numbers could be maintained at a range of low densities. We had neither the space, nor the manpower to carry out one such trial, let alone a factorial combination of trials to investigate the interaction between the duration, timing and intensity of defoliation. The study was restricted, therefore, to a consideration of the effects of the timing and duration of exposure to high densities of rabbits on the survivorship, growth, ear weight and seed weight of winter wheat.

METHODS

The research was carried out in two fields known as Pound Hill and Ashurst Warren on acid, sandy soils of the Bagshot Series at Silwood Park, Berkshire (GR 14 936692). In both experimental fields, following herbicide application in August, liming and ploughing, crops of winter wheat (variety Hustler) were sown in October at a rate of 350 seeds m^{-2} and a spacing of 9 rows per metre. Fertilizer was applied at sowing at a rate of 75 kg ha^{-1} P_2O_5 and K_2O, in March at a rate of 50 kg ha^{-1} N and in late April at a rate of 120 kg ha^{-1} N. Soils were limed each year at a rate of 3 t ha^{-1}. Herbicide treatment of the crop involved a pre-ploughing application of Roundup (4 l glyphosate in 200 l water ha^{-1}), pre-emergence treatment with Chandor (4 l Trifluralin and Linuron in 200 l water ha^{-1}), and a spring application of Agroxone (5 l MCPA in 200 l water ha^{-1}), with spot treatment of *Holcus mollis* with Roundup using a wick applicator.

The field at Pound Hill was exposed to grazing by a natural, free-ranging population of rabbits whose harbourage was in a semi-natural woodland of oak, *Quercus robur*, and sycamore, *Acer pseudoplatanus*, with a dense undergrowth of bramble, *Rubus fruticosus*, and bracken, *Pteridium aquilinum*. The experimental area began some 25 m from the woodland boundary. As soon as the wheat crop grew clear of the soil surface, rabbits were observed in all parts of the field. Rabbits emerged from a 50m length of harbourage to feed, and, while we made no attempt to determine the precise amount of rabbit feeding in each part of the field, up to 10 individuals could be seen on the field at any one time, and it was usual to see between four and six rabbits on the experimental area in the early evening during the winter months (i.e. approximately 3.3 rabbits ha^{-1}, or 0.1 rabbits seen per metre of harbourage). The rabbit numbers observed at Pound Hill would be classed as high by national standards. This, coupled with the small size of our field, and the small individual plot sizes, meant that the rabbit grazing pressures experienced in this study were very high. In 1982–83 there were six different treatments: never grazed, continuously grazed and grazed for 42, 91, 121 and 183 days before fencing. In 1983–84 we added *unfencings* to the design, and allowed the crop to grow for 52, 121 and 198 days before removing the fences and exposing the crop to rabbit grazing for the remainder of the growing season. As in 1982–83, a previously exposed crop was fenced after 52 and 121 days of grazing. In 1984–85 the design was rather different. In addition to never grazed and continuously grazed, we exposed the crop for 30 or 60 day periods and then refenced it. This enabled the importance of the timing of rabbit damage to be determined. The crop was exposed for 30 day periods in November, December,

January, February, March, April and May and for 60 day periods in November and December, January and February, March and April, and May and June.

The rabbit-exclusion plots used at Pound Hill measured 7 m × 3 m and were constructed of 2 cm wire mesh, using a single 10 cm fence post at each corner. The base of the wire was bent outwards for 20 cm and buried in a 5 cm deep trench and the wire was stapled to the posts. Canes were woven through the mesh and pushed firmly into the ground midway down the long side of the plot to improve the rigidity of the fence. We never found signs of rabbits having gained entry to any of the experimental exclosures in any of the three years. Smaller vertebrates were seen (or caught in Longworth traps), including wood mice (*Apodemus sylvaticus*), harvest mice (*Micromys minutus*) and voles (*Microtus agrestis*).

The second experimental site was a plot measuring 60 m × 30 m situated below Ashurst Orchard, Silwood Park. This was fenced to 2 m in height with 2 cm wire mesh, capped by a T-section of wire mesh. The whole fence was then topped by a single strand, electric fence above the T-section to keep out cats and foxes. We established a confined population of rabbits within the fence by introducing two male and two female wild rabbits, trapped just outside the fence. The rabbits were marked with plastic ear tags and released at the beginning of 1982. The rabbits caused very similar levels of damage to the wheat crop to those observed at Pound Hill (i.e. all the plants were grazed virtually to the ground for the entire period from November until April). The rabbits were provided with supplementary food throughout the winter in the form of rabbit pellets and sliced carrot. They bred successfully throughout the course of the experiment and young were removed periodically by live trapping, and released outside the enclosure. Plots were laid out within the area as follows. The field was split into four blocks and, within each block, plots were allocated to five grazing treatments at random. Fences measuring 5 m × 4 m were erected to exclude rabbits at different times after sowing and to allow regrowth of the crop (durations of grazing were 0, 52, 121, 220 and 240 days). Each area was further treated with aphicide and with herbicide in a split plot, factorial design.

RESULTS

The grain yields are shown in Figure 15.1. At Ashurst, yield measured as grain dry weight declined as a roughly linear function of duration of exposure in both years. There was no evidence of compensation for early grazing. At Pound Hill, the trend was again roughly linear, but the pattern was obscured by the fact that we experienced an increasingly severe

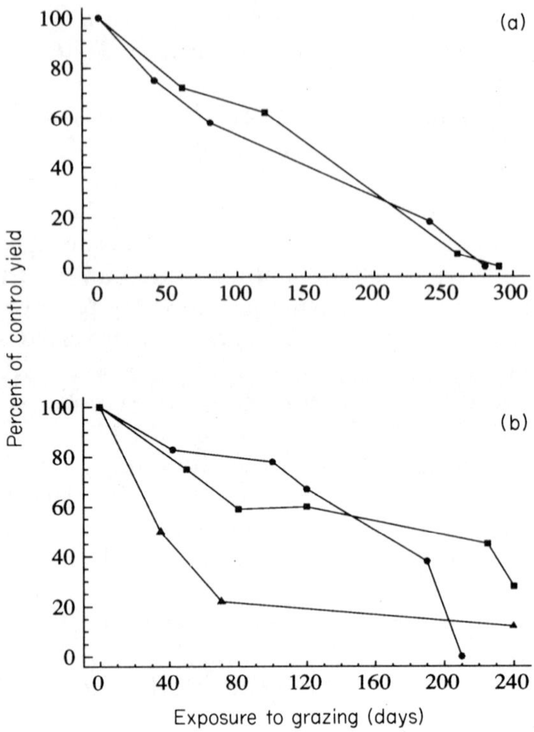

Figure 15.1 The relationship between the duration of exposure to rabbit grazing, measured in days after sowing, and the yield of grain at harvest, expressed as a percentage of control yields, of (a) Ashurst Field in 1983 and 1984, and (b) Pound Hill Field in 1983 (circles), 1984 (squares) and 1985 (diamonds).

problem with infestation of the grazed plots by the perennial grass weed *Holcus mollis* over the three years. The ungrazed wheat crop was able to suppress the weed effectively, but on grazed plots the grass frequently attained dominance. As before, there was no evidence of compensation for early, brief exposure to rabbit grazing.

The effects of timing of exposure are shown in Figure 15.2 for a range of 60-day grazing bouts. Yields suffered most from the earliest defoliation; the longer the onset of grazing was delayed, the lower the losses in yield. Close to harvest time, rabbits entered the crop and felled whole shoots, from which the ears were eaten. There was no time for the crop to recover from this kind of damage by regrowth.

The components of yield loss are shown in Figure 15.3. Grain numbers per ear and individual grain weights suffer severely, even when defoliation occurred several months before grain filling began. Also, the

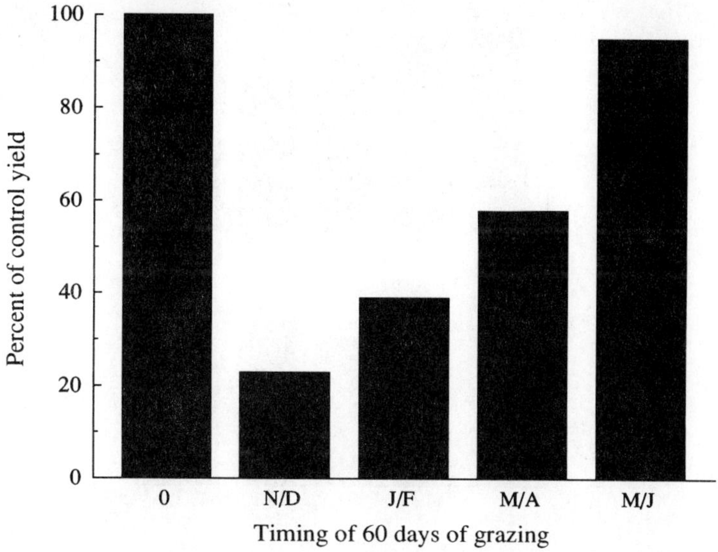

Figure 15.2 The effect of exposing winter wheat to rabbit grazing for 60-day periods at different times of year. Data from Pound Hill Field, 1985, expressed as percentage of control yields.

proportion of the ear made up of chaff increases with rabbit grazing, so yield quality suffers doubly; grain size is reduced so a larger fraction of the crop is lost at threshing, and a larger proportion of ear weight is lost as chaff. Other quality effects are felt through increased pollution of the crop with weed seeds, and uneven ripening (the heavily grazed headlands ripen anything up to two weeks later).

All components of yield were affected by rabbit grazing. Taking 30-day exposure during November 1985 as an example, their effects on yield loss were as follows: (1) reduced numbers of ears per unit area, due to shoot mortality (38.5% loss); (2) reduced ear size (18.4% loss); which in turn is due to (3) reduced numbers of grains per ear (7% loss); and (4) reduced individual mean grain weights (22.3% loss).

In addition to these direct effects of defoliation on cereal yield, there are indirect effects of rabbit grazing that affect yields through increased weediness and increased cereal aphid numbers; these results are shown in Tables 15.1 and 15.2. Table 15.1 gives the grain yields of grazed and ungrazed plots to which additional herbicide was applied during the rapid growth phase. Increased weediness depresses grain yield by an extra 47% on continuously grazed plots. The effect of selective defoliation of the wheat crop is to reduce the competitive ability of the cereal plants relative

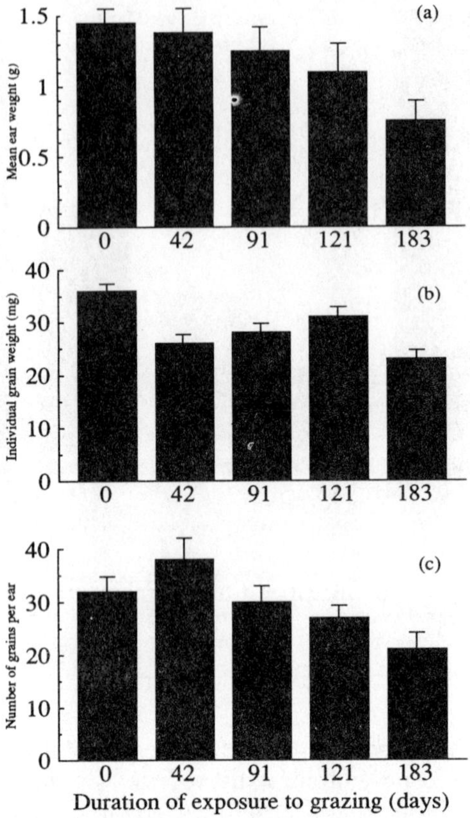

Figure 15.3 Components of yield loss for winter wheat crops grazed for 42, 91, 121 and 183 days after sowing, compared with ungrazed control plants (fenced immediately after sowing and grazed for 0 days). (a) Mean ear dry weight (g); (b) individual grain dry weight (mg), obtained by weighing all the grains from 30 hand-milled ears; (c) numbers of grains per ear (the apparent increase following 42 days exposure is probably an artefact of the chance event of obtaining low grain numbers in the controls). Bars show one standard error.

to less-intensively grazed weed species. In ungrazed plots, the vigorously growing wheat plants act as their own weed killers, suppressing weed growth by the dense shade they cast. In grazed plots, weeds like *Holcus mollis* and *Rumex acetosella* that are avoided by rabbits can grow tall enough to over-top and out-compete the defoliated wheat plants.

Table 15.2 shows the response of grain yield to aphicide application. Aphids were more abundant on the grazed plots because the age structure of the leaf population was dominated by younger, more susceptible

Table 15.1 Grain yields (g dry weight from two rows of length 1 m) from grazed and ungrazed plots treated with an extra application of herbicide during the rapid growth phase. Note the slight yield depression in the ungrazed plots, caused by accidental herbicide carry-over to the crop plants, but the dramatic yield increase on the continuously grazed plots

Exposure (d)	0	52	121	220	240
Weedy	115.9	87.3	68.4	48.4	32.0
Treated	100.8	84.7	81.3	54.8	47.1
% change	−13	−3	19	13	47

Table 15.2 Grain yields from grazed and ungrazed plots treated with aphicide and herbicide. There was no difference in grain yields with and without aphicide on weedy plots (see Table 15.1). Excluding cereal aphids led to significant ($p < 0.05$) yield increases on plots grazed for more than 200 days

Exposure (d)	0	52	121	220	240
Unsprayed	100.8	84.7	81.3	54.8	47.1
Aphicide	102.2	86.3	85.6	60.2	54.2
% change	1	2	5	10	15

foliage. Removing the aphids with pesticide, however, only led to measurable increases in yield on split plots where weed abundance had been reduced by extra herbicide application. On the weedy plots, the increased cereal growth resulting from aphid exclusion was small compared to the reduced performance that came about from competition with those weed species that were avoided by rabbits.

DISCUSSION

We began with the expectation that early, brief exposure to rabbit grazing might lead to negligible yield losses, or perhaps even to yield enhancements, as a result of increased rates of tillering or other forms of compensation (Bardner and Fletcher, 1974; Crawley, 1983). After all, until recently, it was relatively common practice to graze winter cereals in autumn, in order to stop them from becoming 'winter proud'. Our results, however, were contrary to this expectation. In no case was yield compensation observed, and brief early grazing led to eventual yield losses of between 17 and 30% over 3 years at two sites.

Increasing the duration of exposure to rabbit grazing caused a roughly linear decline in grain yields. At one site, yield losses were exacerbated by a progressive increase in the abundance of grass weeds on the grazed plots over the 3 years. The timing of exposure had a substantial effect on yield. For a 60-day bout of feeding, losses were greatest when grazing occurred immediately after germination, in November and December. If grazing exposure lasted only 30 days, then there was less difference between the effects of different timings, with similar losses experienced for all months from November through to March. Exposure for 30 days during the rapid growth phase (April to June) was less damaging, presumably because alternative foods were available at this time, and the tall, often wet crop, was unattractive and relatively impenetrable to the rabbits.

Indirect effects of grazing had important consequences on harvest yield and quality. Increased weediness on grazed plots led to further yield reductions of up to 47%. There were more cereal aphids on the grazed plots, but these only added to yield losses when the extra weeds were experimentally removed. Delay of ripening was patchily distributed within the field, and different parts of the crop matured at different times. Grain quality was reduced by a direct effect of defoliation in reducing seed weight, and indirectly as a result of contamination of the grain by weed seeds.

Even though ear number, ear weight, grain number and individual grain weight were all substantially reduced on grazed plots, some individual plants performed remarkably well, even under continuous grazing. There appears to be considerable scope for experimental manipulation of cultural practices (sowing densities, rates and timing of nitrogen application, etc.) aimed at maximizing the proportion of the plants that regrow to their maximum capability. It is unlikely, however, that any amount of cultural manipulation will cause these modern varieties of winter wheat to over-compensate, in such a way that grazed plots out-yield their ungrazed counterparts.

In arable fields infested by pest populations of rabbits, it is unlikely that individual plants experience 'low intensity' defoliation; they are either grazed to the ground, or they are not grazed at all. While it is clear that the crop might compensate for this kind of damage by gap-filling (Bardner and Fletcher, 1974), there is no evidence that the total crop yield could be increased by this means. If lower density crops produced higher yields, then farmers would, presumably, sow fewer seeds.

In quantifying the effects of both the timing and the duration of exposure to rabbit grazing, we have fulfilled two of our three aims. We are unable to carry out experiments on the effects of different intensities of grazing, however, and this remains an important challenge for future studies. These experiments will not be straightforward, however, because in order to obtain very low grazing pressures, such large enclosures will be needed that the plots are almost certain to be internally heterogeneous (or to become heterogeneous as a result of patchy grazing). This will mean that obtaining adequate replication is difficult. Nevertheless, the protocols now exist for assessing the impact of rabbit grazing on cereal yields, and these experiments are perfectly feasible. In addition to the simple expedient of rabbit-proof fencing, there are clear signs that yield losses could be reduced, and profitability increased, by judicious use of extra herbicide and probably by a range of other cultural practices.

ACKNOWLEDGEMENTS

This study was carried out under contract to the Chief Scientist's Group of MAFF. Many thanks to Richard Brown who worked on this contract for the first two years, and to the staff of The Rabbit Group, Tangley Place, Worplesdon, for their help throughout.

REFERENCES

Bardner, R. and Fletcher, K.E. (1974) Insect infestations and their effects on the growth and yield of field crops: a review. *Bull. Entomol. Res.* **64**, 1–11.

Crawley, M.J. (1983) *Herbivory: The Dynamics of Animal-Plant Interactions.* Blackwell Scientific Publications, Oxford.

Crawley, M.J. (1988) Rabbit grazing, plant competition and seedling recruitment in acid grassland. *J. Ecol.* (submitted).

Crawley, M.J. and Brown, R.A. (1988a) The impact of rabbit grazing on the yield of winter wheat. *J. Appl. Ecol.* (submitted).

Crawley, M.J. and Brown, R.A. (1988b) The impact of rabbit grazing on the growth and dry matter distribution of winter wheat. *J. Appl. Ecol.* (submitted).

Mills, S. (1986) Rabbits breed a growing controversy. *New Scientist*, 6 March, 50–4.

Sixteen

Rabbit ranging behaviour and its implications for the management of rabbit populations

D.P. COWAN, A.R. HARDY, J.P. VAUGHAN AND W.G. CHRISTIE

SUMMARY

The ranging behaviour of wild rabbits using a 6 ha field of winter barley was studied for two consecutive winters by direct observation and radio-tracking techniques. The frequency with which a given rabbit used the field was influenced by the distance of its daytime refuge from both this field and alternative fields, and whether the refuge was above or below ground. Rabbit activity across the study field declined with increasing distance from the field margin. The proportion of the time spent grazing initially increased and then decreased with distance into the field. The implications of these results are discussed in relation to the management of rabbits in order to reduce crop damage.

INTRODUCTION

The national population of the European wild rabbit (*Oryctolagus cuniculus*) has increased steadily since the early 1950s when myxomatosis was estimated to have killed 99% of the then population (Ross, 1982; Trout, Tapper and Harradine, 1986). Rabbits are once again considered to be a major pest of agricultural crops (Rees *et al.*, 1985). Effective methods to control rabbits and to reduce crop damage are not universally applicable throughout the diversity of agricultural habitats. Research is therefore underway to quantify crop losses, assess the sizes of rabbit populations and to develop better management techniques. Since successful management exploits the biology of a pest species, the way in which rabbits use space in an agricultural habitat has clear practical implications. Although detailed studies of rabbit ranging behaviour have been carried out in some habitats, notably chalk grassland and sand-dune systems (Cowan and Garson, 1985; Cowan, 1987), less is known about rabbit movements on agricultural land. A study was therefore undertaken to relate rabbit movements and grazing activity to plant growth and crop yield in a field of winter cereals. This chapter summarizes the ranging behaviour of rabbits while the crop aspects are dealt with in detail by Crawley (Chapter 15).

METHODS

Rabbits were observed over two consecutive winters on a 6.5 ha field of winter barley (variety 'Igri') in Hampshire which was bounded by a canal and bridleway on two adjacent sides which did not provide refuges. Rabbits moved onto the study field from the other two sides which were formed by a hedgerow beyond which were two grass paddocks and a 5 ha wood.

Rabbits, caught in baited cage-traps on the field, were fitted with radio-collars and ear tags with an individual combination of beta lights and scotchlite reflective tape. Radio equipment consisted of transmitters on 104 MHz, Yaesu receivers and H-adcock aerials. Activity data were collected in four ways.

1. Radio-tagged rabbits were located every two days to determine the location of their daytime refuges. Distances from these refuges to the study and alternative fields were measured.
2. Nocturnal activity on the field was observed from an elevated hide opposite the hedgerow and woodland margins. Fixes were regularly taken to determine the frequency with which radio-tagged rabbits were present on the field.
3. Using an image intensifier and a laser torch, repeated scans of the field were made to note the position of all rabbits present. Animals were assigned to $5\,m \times 5\,m$ grid squares within a $10\,m \times 10\,m$ grid of reflective tape markers laid out on the field. In order to reduce observer

bias and errors in ascribing rabbits to grid squares, a smoothing algorithm was used:

$$Y=[X+0.5(X1+X2+X3+X4)+0.25(X5+X6+X7+X8)]\ a/100b$$

where Y is the number of rabbit minutes per day per m^2, X is the number of sightings in the grid square concerned, $X1$, $X2$, $X3$ and $X4$ are the numbers of sightings in the four laterally adjacent grid squares, $X5$, $X6$, $X7$ and $X8$ are the numbers of sightings in the four diagonal adjacent grid squares, a is the number of minutes per day (using a 720 minute 'day') that rabbits were active on the field and b is the number of scans made during the period concerned.

4. Continuous 15 minute records were made of the behaviour of individual rabbits chosen at random from those present on the field at any one time.

The study was conducted from September 1982 to April 1984. During the 1982/3 winter observations were made throughout the period between dusk and dawn. However, a constant level of activity was found between 1900 and 0500 h with indications of a slight peak in activity between 2100 and 2200 h. On the basis of this analysis, observations were confined to the period 1900 to 2300 h during the second winter period. The two barley crops appeared very similar in the two years with only one week's difference in sowing dates.

Data were analysed to address two basic questions. What factors determine the frequency of use of a field by a given rabbit? What factors influence the distribution and behaviour of rabbits within a field?

RESULTS

A total of 24 rabbits were marked during the study and data collected on 26 individuals: 18 males and 8 females (two rabbits were observed in both winters). The frequencies with which individual rabbits visited the study field were investigated by correlating the proportion of radio-tracking fixes that were on the study field with a number of parameters (Table 16.1). The best fit multiple regression models of the relationship between the proportion of fixes on the study field and these parameters are shown in Table 16.1. For the purposes of this analysis, the study period was divided into four – September to December 1982, January to March 1983, September to December 1983 and January to April 1984. There were no differences between these periods ($F=0.039$, df 3, 2 NS) or between sexes ($F=1.139$, df 1, 22 NS).

Table 16.1 Correlation coefficients between the arcsine transformed proportion (arcsprop = arcsine $\sqrt{1-P}$) of radio-tracking fixes (minimum number = 9, $\bar{x} = 45.2 \pm \text{SD}37.8$) and the mean distance (m) of the rabbits' daytime refuges (minimum number of fixes = 5, $\bar{x} = 17.3 \pm \text{SD}8.7$) from the study field (dist stud), from the nearest alternative field (dist alt), the mean difference between the distances of the study field and the nearest alternative field from daytime refuges (dist stud − dist alt = diff), the proportion of daytime refuges that were underground (day und) and the proportion of daytime refuges that were at the same position as the previous daytime location (prop sam):

	dist stud	dist alt	diff	day und	prop sam
arcs prop	−0.48**	0.16	−0.48**	0.44**	0.39*

(Significance levels: ** $p<0.01$, * $p<0.05$).
The best fit multiple regressions derived were:
(a) including proportion underground (prop und) and prop sam

$$\text{Arcs prop} = 0.37 - 0.00252\ \text{diff} + 0.29\ \text{prop sam} - 0.00072\ \text{dist alt} - 0.018\ \text{prop und}\ (R^2 = 0.42)$$

(b) excluding prop und and prop sam

$$\text{Arcs prop} = 0.54 - 0.00324\ \text{diff} - 0.00191\ \text{dist alt}\ (R^2 = 0.33)$$

From a practical point of view, the most important factor determining the frequency with which rabbits travelled to feed on the study field was the distance of their daytime refuge from this field. Rather than absolute distance, the difference in distance between the study and alternative fields was important. If a rabbit's daytime refuge was more than 100 m closer to the alternative field then it rarely visited the study field. In addition, if a rabbit's refuges were predominantly underground in burrow systems rather than above ground in bramble thickets, or in shallow forms, then it visited the study field more frequently, perhaps reflecting less use of the food available within the woodland and hedgerow refuge locations. Similarly individuals who rarely changed refuge sites were more frequent visitors to the study field than those who frequently moved refuge locations.

Regarding the distribution and behaviour of rabbits within the field, two factors emerged of particular importance. First the distance from cover at the field edge. Intensity of use declined as an approximately exponential function of distance from cover, such that only 5% of rabbit sightings were at distances of more than 40 m from the field edge (Figure 16.1). Comparison between years revealed a significant difference which may have been due to increased wariness after a year of intensive study on the field. There were also changes in behaviour with increasing distance from cover (Figure 16.2). At the field edge the proportion of time rabbits spent feeding was relatively low and this was matched by a relatively high

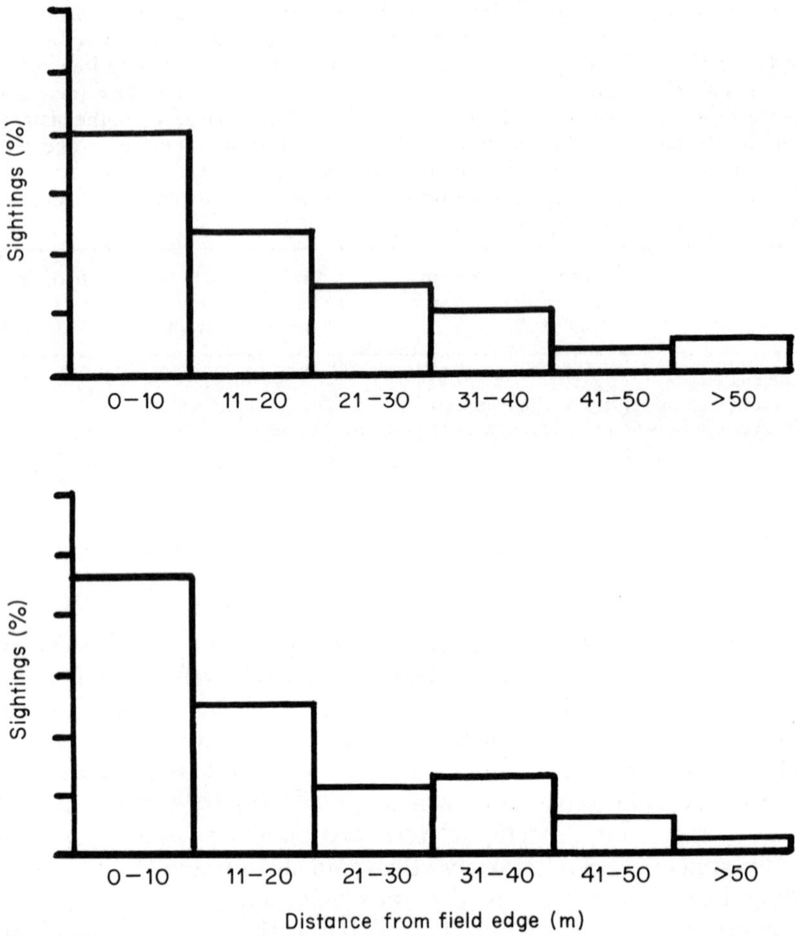

Figure 16.1 The nocturnal distribution of rabbits from the edge of a field of winter barley estimated during scans from an elevated hide: (a) winter 1982/83 where percentage of sightings = 234 × distance from field edge$^{-0.93}$, $r = 0.98$, $p < 0.05$); (b) winter 1983/84 where percentage of sightings = 138 × distance from field edge$^{-0.74}$, $r = 0.92$, $p < 0.1$).

proportion of time spent vigilant. The proportion of time spent feeding increased for rabbits up to 30 m from cover and declined again for rabbits 35 m or more from cover, matched by an increase in vigilance. Such behavioural differences may have reflected changes in predation risk with increasing distance from cover. Some rabbits were active on the field throughout the night; however, on average, only one-third of the rabbits

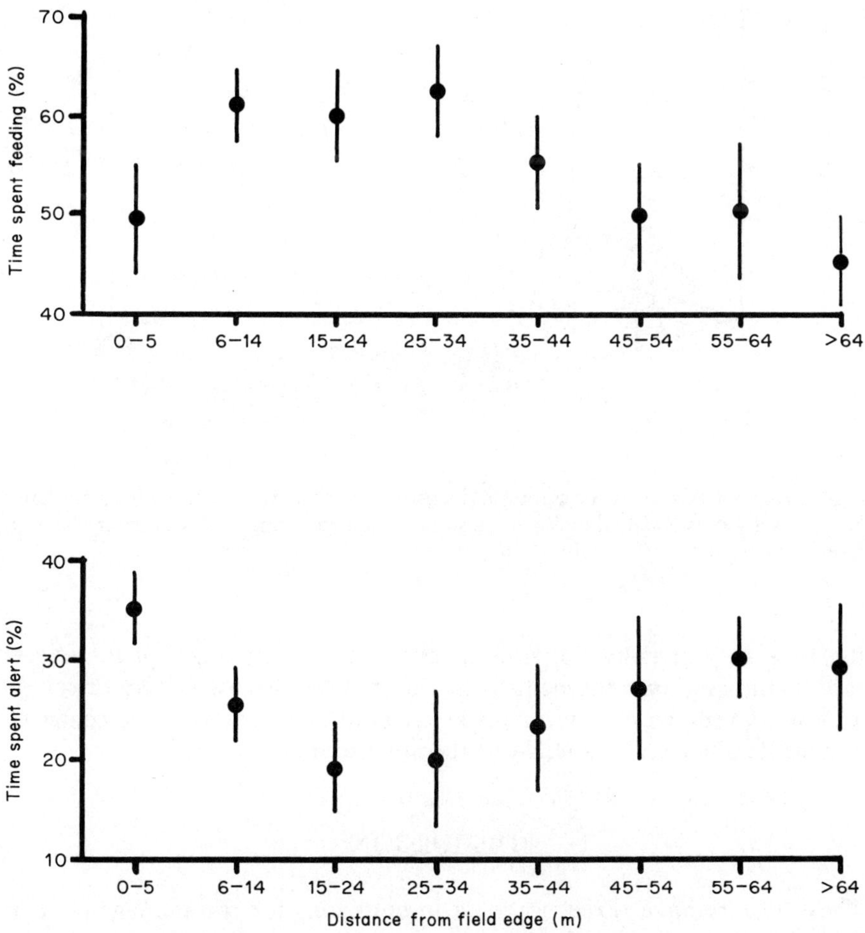

Figure 16.2 Variations in the percentage of time spent by rabbits on a field of winter barley either grazing ($F = 6.84$, df 7971, $p < 0.0001$) or alert ($F = 4.98$, df 7971, $p < 0.001$) at different distances from the field edge. Behavioural observations were made during nocturnal watches from an elevated hide during two consecutive winters.

using the field were present at any one time. There were five-fold differences in the sizes of individual ranges on the study field.

The other factor found to influence the distribution of rabbits on the field was the distribution of their daytime refuges around the field edge. Figure 16.3 is a three-dimensional plot of average rabbit activity across

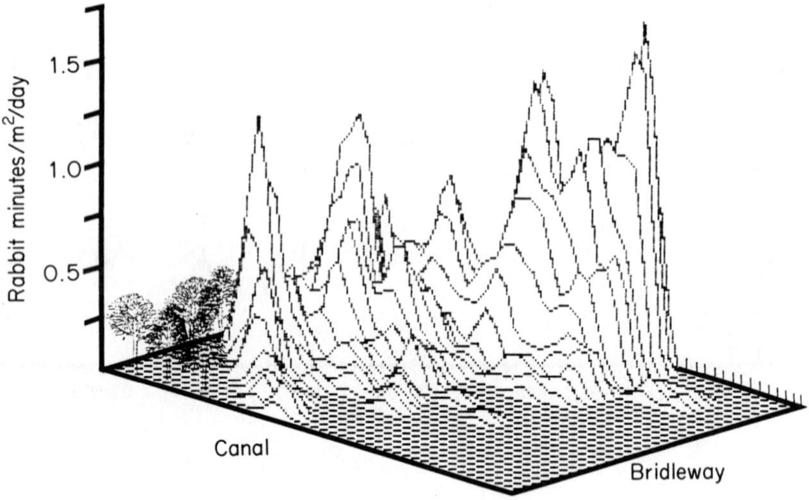

Figure 16.3 Three-dimensional plot of nocturnal rabbit activity across a field of winter barley. Activity was calculated using a smoothing algorithm, as described in the text, from observations made from an elevated hide over two winters.

the field in which five obvious peaks can be seen. Each peak reflects rabbits moving onto the field from a particular burrow system or refuge area within the woodland or hedgerow and may represent the centre of activity of a particular social or breeding group.

DISCUSSION

These results have three practical implications for the management of rabbits to reduce crop damage. First, in terms of determining the size of a problem population, the observations of distances that rabbits move to feed on a given field provide insight into the area over which management techniques must be applied for them to be fully effective. Secondly, the measurements of the distribution of rabbit activity across the field assist the assessment of the area over which crop loss is likely and the variation in its extent. Finally, the observations of rabbit behaviour reported here help to identify the optimum distribution within a field of specific management techniques using baits, for instance, the positioning of baited cage traps. Generalizations based on a single study must of course be tentative. However, this study does point the way for more extensive investigations elsewhere.

REFERENCES

Cowan, D.P. (1987) Aspects of the social organisation of the European wild rabbit (*Oryctolagus cuniculus*). *Ethology,* **75**, 197–210.

Cowan, D.P. and Garson, P.J. (1985) Variations in the social structure of rabbit populations: causes and demographic consequences in *Behavioural Ecology: Ecological Consequences of Adaptive Behaviour* (eds R.M. Sibly and R.H. Smith) Blackwell Scientific Publications, Oxford, pp. 537–55.

Rees, W.A., Ross, J., Cowan, D.P. Tittensor, A.M. and Trout, R.C. (1985) Humane control of rabbits in *Humane Control of Land Mammals and Birds* (ed. D.P. Britt), The Universities Federation for Animal Welfare, Hertfordshire, pp. 96–102.

Ross, J. (1982) Myxomatosis: the natural evolution of the disease. *Symp. Zool. Soc. Lond.,* **50**, 77–95.

Trout, R.C., Tapper, S.C. and Harradine, J. (1986) Recent trends in the rabbit population in Britain. *Mamm. Rev.,* **16**, 117–23.

Seventeen

Population dynamics of parasites of the wild rabbit

B. BOAG

SUMMARY

The influence and potential exploitation of factors which may naturally control vertebrate pests has, apart from myxomatosis, received very little attention in Great Britain. This chapter presents the results of a 10 year study of the population dynamics of three nematode and two cestode species infecting wild rabbits and the prevalence and severity of symptoms of the protozoan parasite *Eimeria stiedae* damage to rabbit livers. The stomach nematode *Graphidium strigosum* was more prevalent, and infections were more intense in adult rabbits in early spring whereas the seasonal fluctuations of the small intestine nematode, *Trichostrongylus retortaeformis* were less marked, the intensity of infection being greatest in young animals during the autumn. Fewer rabbits were infected with the tape worm *Mosgovoyia pectinata* and the large intestine nematode *Passalurus ambiguus*, both the percentage of animals infected and the size of the infections being lower in the late spring and summer than during the rest of the year. The tape worm *Cittotaenia denticulata* was more prevalent during the autumn whereas populations tended to be larger in mid-summer. Symptoms of *Eimeria stiedae* damage to rabbit livers were most prevalent between June and September, the most severe symptoms being recorded from young rabbits weighing 500–1000 g. The role of parasites in restricting the size of rabbit populations is reviewed.

INTRODUCTION

The rabbit is probably the most important economic pest occurring in Great Britain and yet there are many aspects of its biology and ecology of which very little is known. Although myxomatosis dramatically reduced numbers in the late 1950s (to perhaps 1% of their former levels) numbers have begun to increase again in recent years and in spite of the use of other control measures, e.g. gassing, shooting, snaring and habitat management, and the testing of new control measures e.g. acoustic scaring devices and warren ripping, continue to increase.

Surveys from throughout the United Kingdom have shown rabbits are infected with both ecto- and endoparasites which may cause rabbits to lose condition, and may thus influence population performance and of course the rabbit flea, *Spilopsyllus cuniculi*, acts as the vector for the myxoma virus. The present chapter is concerned with only the endoparasites. Nematodes, cestodes, a trematode and protozoan parasites (Mead-Briggs and Page, 1975; Boag, 1985, Cowan, 1983) have been recorded in Great Britain but their influence on the growth, fecundity and survival of rabbits has not been studied. Investigations into the relationship between rabbits and their parasites could lead to both a better understanding of the population dynamics of wild rabbit populations and may assist in the formulation of novel management procedures. Dunsmore (1980) suggested the population productivity of rabbits in Australia was increased because of the reduced parasite fauna in that country compared with rabbits in Europe. He also demonstrated experimentally that *Trichostrongylus retortaeformis* could reduce the number of kittens weaned by 20% and that kittens from worm-free does were 13% heavier. *T. retortaeformis* has been shown to damage the mucosa of the small intestine (Barker and Ford, 1975) and in extreme cases when populations are very large, *T. retoraeformis* has been shown to kill animals (Michael, 1952). *Graphidium strigosum* has been implicated with the death of hares (Broekhuizen and Kemmers, 1976) and Bull (1953) suggested that heavy infections in the rabbit might weaken the stomach wall and lead to a rupture of the stomach and death of the rabbit. Enigk (1938) found severely infected rabbits on reduced rations lost more weight than uninfected rabbits. However, after a 10 year study of the ecology of the wild rabbit, Bull (1964) concluded that *G. strigosum* had little or no adverse effect on the weight or fertility of rabbits in New Zealand, and, in the absence of experimental evidence the role of the nematode in regulating rabbit populations in Great Britain must be open to question.

To ascertain the role that helminth parasites may have in restricting the size of rabbit populations it is necessary to know their population dynamics. Investigation in the United Kingdom (Mead-Briggs and Vaughan, 1973; Boag, 1985), Australia (Mykytowycz, 1956) and New Zealand (Bull, 1964) have shown marked seasonal differences in the seasonal prevalence and incidence of the various helminth species. Boag (1988) has also found an interaction between some of the helminth species and myxomatosis.

The identification of protozoan parasites is more difficult than that of the helminth parasites and consequently less is known about their distribution, prevalence or the intensity of naturally occurring infections. However, may protozoan parasites belonging to the genus *Eimeria* have been recorded from rabbits from Great Britain (Catchpole and Norton, 1979; Cowan, 1983), Australia (Mykytowycz, 1956; Stodart, 1971) and New Zealand (Bull, 1958). The pathogenicity of some of these protozoan parasites is well documented. *Eimeria stiedae* which invades the liver has been shown by field observations (Tyndale-Biscoe and Williams, 1955; Bull, 1958; Stodart, 1971) to be relatively common and laboratory experiments (Barriga and Arnoni, 1979, 1981; Patterson, 1987) and trials with coccidiostats (Peeters *et al.*, 1981) have shown it to be particularly pathogenic. Whittle (1955) even suggested this species could be responsible for a 10 to 15 year cycle in rabbit numbers. Some intestinal *Eimeria* species have been shown to be pathogenic under artificial conditions but Patterson (1987) suggests that for domestic rabbits 'the intestinal coccidiosis of clinical significance, without the participation of other agents (i.e. bacteria) is relatively rare'.

Studies on the population dynamics of *Eimeria* species have usually been confined to results obtained from faecal samples but post-mortem findings from work done in Australia and New Zealand have also shown that acute disease, especially due to *E. stiedae*, is usually confined to young rabbits raised in densely inhabited colonies which have access to and graze damp, lush vegetation (Tyndale-Biscoe and Williams, 1955; Bull, 1958; Mykytowycz, 1962; Stodart, 1971; Cowan, 1983).

The aim of the present work was to investigate the population dynamics of the more common parasites of the wild rabbit and identify when these parasites are most numerous and likely to influence the growth, fecundity and survival of their host.

MATERIALS AND METHODS

Rabbits were collected monthly (with the exception of November 1978 and April 1979) between January 1977 and December 1986 from Littleton Farm, Inchture, Perthshire, Scotland. The rabbits were shot using a rifle, which gave a relatively unbiased sample (Boag, 1987) and left the viscera intact. In the laboratory the rabbits were weighed, sexed and the abdominal cavity opened and the contents removed. The tract was separated into the three regions, stomach, small intestine and large intestine before being opened along its entire length and the contents sieved through a 100 mesh (125 μm) sieve. The residues left on the sieves were collected and either examined directly or stored in 5% formalin. If the number of nematodes in a sample was less than 25 all were counted, while larger populations were examined by counting a known dilution of the total sample. The maximum dilution never exceeded one in 25. Cestodes were counted, prepared and identified as described by Boag (1985). The livers were examined for the presence of lesions caused by *E. stiedae* and severity of symptoms assessed according to the scale: none, (no lesions found); light (less than 5 spots present); moderate (5–25 lesions); heavy (up to half the area of liver affected by lesions); very heavy (more than half the liver affected by lesions).

RESULTS

The prevalence and incidence of the five most commonly occurring helminths can be seen in Figure 17.1. Over 50% of the rabbits were infected with *G. strigosum* between November and April while greater numbers were recorded in adult animals between February and March. In many of these animals the stomach wall was found to be very thin and transparent and an estimate of the worm burden would have been possible without opening and removing the stomach contents. *Trichostrongylus retortaeformis* was the most prevalent species overall but, unlike *G. strigosum* its incidence was greater in juvenile rabbits in September and October. The size of *P. ambiguus* populations differed considerably between rabbits collected during any one month but there was a trend towards higher populations in adult rabbits in mid-winter. The population dynamics of the two cestode species were dissimilar. *Mosgovoyia pectinata* was found most often in adult rabbits over the winter period and populations were greater in late autumn and early winter, whereas *C. denticulata* populations were greatest in mid-summer and it was most prevalent in late autumn.

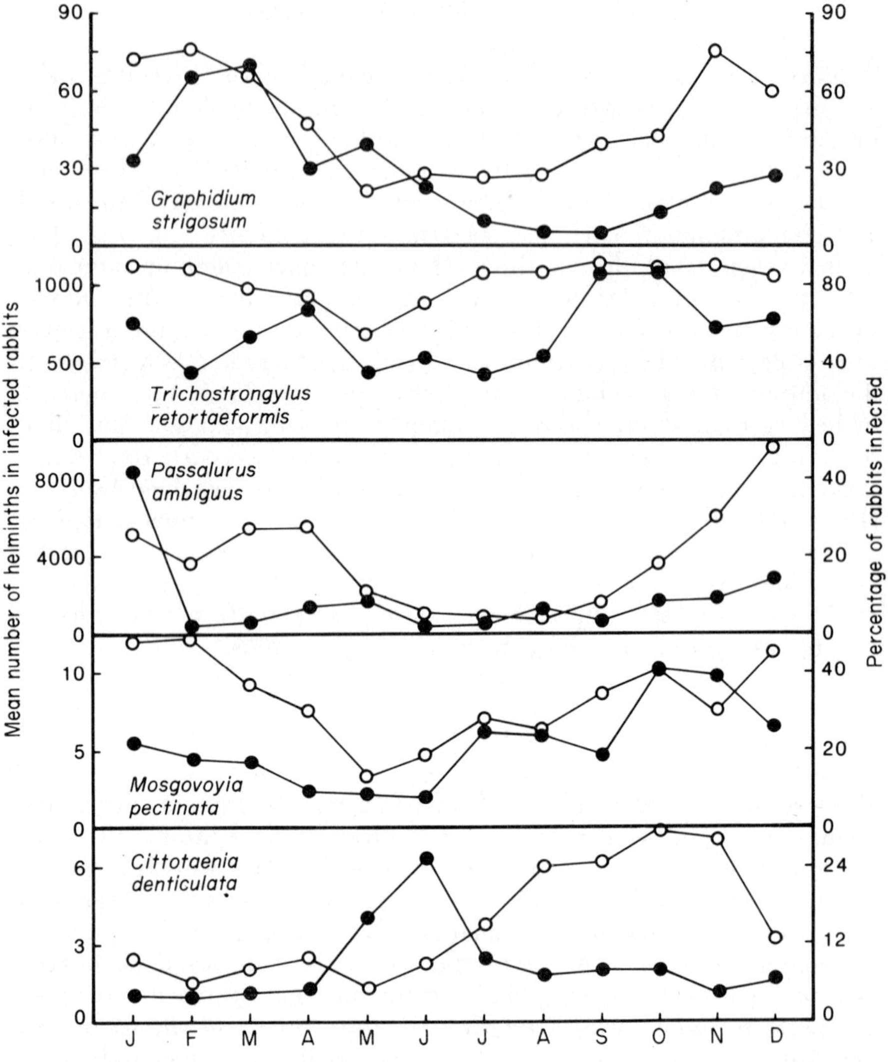

Figure 17.1 The seasonal variation in prevalence (%) ○———○ and intensity ●———● of parasite infections.

The prevalence and incidence of *E. stiedae* symptoms in rabbit livers can be seen in Tables 17.1 and 17.2. No symptoms were observed in animals below 250 g whereas over 50% of those weighing 501–1000 g were infected, 20% having heavy or very heavy infections. Low or moderate levels of liver coccidiosis were found throughout the year but

Table 17.1 Relation between body weight and percentage of rabbit livers with symptoms of *Eimeria stiedae*

Weight (g)	Number of livers examined	Severity of infection None	Low	Moderate	Heavy	Very heavy
<250	16	100.0	0	0	0	0
251–500	201	88.5	7.5	1.5	0	2.5
501–750	176	44.9	21.6	10.2	10.2	13.1
751–1000	173	44.5	20.8	12.1	11.0	11.6
1001–1250	176	65.4	17.6	8.5	5.1	3.4
1251–1500	230	81.8	13.0	2.6	1.7	0.9
1501–1750	275	81.4	16.0	0.4	0.7	1.5
>1751	94	82.9	16.0	1.1	0	0
Total	1341					

the most severe symptoms were usually seen in young rabbits between June and October when 5–10% of the entire population had very heavy infections.

DISCUSSION

The harmful effect of helminth and protozoan parasites to domestic animals is well documented (Holmes, 1987; Rommel, 1987) but, with one exception (Dunsmore, 1980) no experiments have been conducted to ascertain the effect of helminths on the wild rabbit. In Great Britain surveys have shown rabbits can harbour at least 10 helminth and 11 protozoan parasites but many are found only rarely and are unlikely to have any significant impact on rabbit numbers (Patterson, 1987). The numbers of *G. strigosum* recovered during the present survey were similar to those reported by Bull (1964) when he concluded that this nematode probably had little or no adverse effect on the weight or fertility of rabbits. However, some rabbits collected during the present investigation might have subsequently died from their stomachs rupturing since the nematodes could be observed, counted and sexed through the very thin remaining stomach wall. The reason why *G. strigosum* numbers peak in adult rabbits in early spring is unknown but this result reflects similar trends recorded from New Zealand (Bull, 1964) and Australia (Dunsmore, 1980).

The *T. retortaeformis* populations found were within the range found by Dunsmore (1980) to adversely affect the reproduction and growth of rabbits and probably had a similar harmful effect on the rabbits collected

Table 17.2 Seasonal variation in the percentage of livers of young rabbits (<1000 g) with differing severity of *Eimeria stiedae* infections

Month	Number of rabbits below 1000 g examined	Severity of infection					Percentage of total population which were below 1000 g
		None	Low	Moderate	Heavy	Very heavy	
January	1	100	0	0	0	0	2
February	0	—	—	—	—	—	0
March	4	100	0	0	0	0	8
April	19	100	0	0	0	0	22
May	107	80	11	5	4	0	64
June	145	64	13	10	8	5	73
July	158	46	21	10	8	16	61
August	77	61	17	8	5	9	49
September	34	55	24	0	12	9	31
October	9	33	11	0	11	45	10
November	4	100	0	0	0	0	7
December	4	75	0	0	0	25	6
Total	562						

during the present survey. Populations tended to be greater in young animals in September and October before the rabbits become adult and their immune response develops and eliminates most of the nematodes. *Passalurus ambiguus* was often found in very large numbers in mid-winter but these rabbits were in good condition and showed no signs of being ill. Neither of the cestodes *M. pectinata* or *C. denticulata* have been shown to have any detectable detrimental effect on rabbits even though they are relatively large parasites which are found throughout Great Britain (Mead-Briggs and Page, 1975). The differences in the population dynamics of the two cestode species could possibly be due to the seasonal availability of the orbatid mites which act as intermediate hosts for these tapeworms (Evans, 1940).

Coccidial parasites, particularly *E. stiedae,* have been cited as potentially important mortality factors controlling rabbit populations (Bull, 1958; Mykytowycz, 1962; Cowan, 1983). The results of the present study found *E. stiedae* occurring in large numbers of young rabbits many of which had enlarged livers (heavy and very heavy infections) and could possibly die. Those with lighter infections may also suffer a reduction in weight gain before the host's age or immune response eliminated the infection (Gomey-Bautista, Rojo-Vazquez and Alunda, 1987). Very young animals have been shown to have a certain level of resistance to infection (Durr and Pellerdy, 1969) and this may account for none of the kittens below 250 g having any symptoms of hepatic coccidiosis. The population dynamics of *E. stiedae* in the present study were similar to previous investigations which showed the acute disease was confined to animals between 3 and 12 weeks old (Bull, 1958; Mykytowycz, 1956).

It is apparent from the literature and from the results reported here that rabbits are subjected to a series of parasitic challenges throughout their lives and that these probably act as imperfectly density-dependent factors contributing to the natural control of rabbit numbers. The effect of the individual parasitic species on the growth, fecundity and survival of wild rabbits is largely unknown but it is possible to make estimates by extrapolating from laboratory experiments or from work on other animals. Young rabbits 3–12 weeks old are particularly at risk from coccidial parasites and some pathogenic species, e.g. *E. stiedae*, are likely to kill or debilitate over 10% of the animals (heavily and very heavily infected) between June and October. *Trichostrongylus retortaeformis* is present throughout the year and probably acts as a chronic disease in rabbits. However, young rabbits in late autumn would seem to be particularly susceptible to large infections and this may reduce their rate of growth, condition and consequently their chances of surviving the winter. Any adverse effect due to *G. strigosum* will probably be confined to adult animals with large infections in early spring. This parasite may kill rabbits or contribute to the loss of condition of adult and suckling animals if there is a shortage of food. There is no evidence to suggest that *P. ambiguus* or the two tapeworms have any detectable harmful effect on rabbit populations.

REFERENCES

Barker, I.K. and Ford, G.E. (1975) Development and distribution of atrophic enteritis in the small intestine of rabbits infected with *Trichostrongylus retortaeformis*. *J. Comp. Pathol.*, **85**, 427–35.

Barriga, O.O. and Arnoni, J.V. (1979) *Eimeria stiedae*: Weight, oocyst output and hepatic function of rabbits with graded infections. *Exp. Parasitol.*, **48**, 407–14.

Barriga, O.O. and Arnoni, J.V. (1981) Pathophysiology of hepatic coccidiosis in rabbits. *Vet. Parasitol.*, **8**, 201–10.

Boag, B. (1985) The incidence of helminth parasites from the wild rabbit *Oryctolagus cuniculus* (L.) in Eastern Scotland. *J. Helminthol.*, **59**, 61–9.

Boag, B. (1987) Reduction in numbers of the wild rabbit (*Oryctolagus cuniculus*) due to changes in agricultural practices and land use. *Crop Protection*, **6**, 347–51.

Boag, B. (1988) Observations on the seasonal incidence of myxomatosis and its interactions with helminth parasites in the European rabbit (*Oryctolagus cuniculus*). *J. Wildl. Dis.*, **24**, 450–5.

Boag, B. and Thomas, R.J. (1975) The population dynamics of nematode parasites of sheep in Northern England. *Res. Vet. Sci.*, **19**, 293–5.

Broekhuizen, S. and Kemmers, R. (1976) The stomach worm, *Graphidium strigosum* (Dufardin) Railliet and Henry, in the European hare *Lepus europaeus* Pallas in *Proceedings of the International Symposium on the Ecology and Management of European Hare Populations, Warsaw* (eds Z. Pielowski and Z. Pucek), Polish Hunting Association, Warsaw, pp. 157–77.

Bull, P.C. (1953) Parasites of the wild rabbit *Oryctolagus cuniculus* (L.) in New Zealand. *N. Z. J. Sci. Technol.* **B34**, 341–72.

Bull, P.C. (1958) Incidence of coccidia (Sporozoa) in wild rabbits, *Oryctolagus cuniculus* (L.) in Hawke's Bay, New Zealand. *N. Z. J. Sci.*, **1**, 289–328.

Bull, P.C. (1964) Ecology of helminth parasites of the wild rabbit *Oryctolagus cuniculus* (L.) in New Zealand. *N. Z. Dep. Sci. Industr. Res. Bull.*, **158**, 1–147.

Catchpole, J. and Norton, C.C. (1979) The species of *Eimeria* in rabbits for meat production in Britain. *Parasitology*, **79**, 249–57.

Cowan, D.P. (1983) Aspects of the behavioural ecology of a free-living population of the European wild rabbit, *Oryctolagus cuniculus* L. in Southern England. PhD Thesis Royal Holloway College, University of London.

Dunsmore, J.D. (1980) The role of parasites in population regulation of the European rabbit, *Oryctolagus cuniculus*, in Australia. *Proc. Worldwide Furbearer Conf.*, **No. 2, 654–69.**

Durr, U. and Pellerdy, L. (1969) The susceptibility of suckling rabbits to infection with coccidia. *Acta Vet. Acad. Sci. Hung., Tomus*, **19**, 453–62.

Enigk, K. (1938) Ein Beitrag zur Physiologie und zum Wirt-Parasit-Verhaltnis von *Graphidium strigosum* (Trichostrongylidae, Nematoda). *Z. Parasit.*, **10**, 657–721.

Evans, N.M.R. (1940) Observations upon some common cestode parasites of the wild rabbit *Oryctolagus cuniculus*. *Parasitology*, **32**, 78–90.

Gomey-Bautista, M., Rojo-Vazquez, F.A. and Alundra, J.M. (1987) The effect of the host's age on the pathology of *Eimeria stiedae* infection in rabbits. *Vet. Parasitol.*, **24**, 47–57.

Holmes, P.H. (1987) Pathophysiology of nematode infections. *Int. J. Parasitol.*, **17**, 443–51.

Mead-Briggs, A.R. and Page, R.J.C. (1975) Records of anoplocephaline cestodes from wild rabbits and hares collected throughout Great Britain. *J. Helminthol.*, **49**, 49–56.

Mead-Briggs, A.R. and Vaughan, J.A. (1973) The incidence of anoplocephaline cestodes in a population of rabbits in Surrey, England. *Parasitology*, **67**, 351–64.

Michael, J.F. (1952) Self-cure in infections of *Trichostrongylus retortaeformis* and its causation. *Nature*, **169**, 881.

Mykytowycz, R. (1956) A survey of endoparasites of the wild rabbit *Oryctolagus cuniculus* (L.), in Australia. *CSIRO Wildl. Res.*, **1**, 19–25.

Mykytowycz, R. (1962) Epidemiology of coccidiosis (*Eimeria* spp.) in an experimental population of the Australian wild rabbit, *Oryctolagus cuniculus* (L.). *Parasitology*, **52**, 375–95.

Patterson, L.T. (1987) Rabbit coccidiosis. *Vet. Hum. Toxicol.*, **29**, 73–9.

Peeters, J.E., Charlier, G., Antoine, O. and Mammerickx, M. (1984) Clinical and pathological changes after *Eimeria intestinalis* infection in rabbits. *Zentral. Veterinarmed.*, **B31**, 9–24.

Peeters, J.E., Geeroms, R., Froyman, R. and Halen, P. (1981) Coccidiosis in rabbits: a field study. *Res. Vet. Sci.*, **30**, 328–34.

Rommel, M. (1987) Control of non-vector-borne protozoan. *Int. J. Parasitol.*, **17**, 639–47.

Stodart, E. (1971) Coccidiosis in wild rabbits, *Oryctolagus cuniculus* (L.), at a site on the coastal plain in Eastern Australia. *Aust. J. Zool.*, **19**, 287–92.

Tyndale-Biscoe, C.H. and Williams, R.M. (1955) A study of natural mortality in a wild population of the rabbit, *Oryctolagus cuniculus* (L.). *N. Z. J. Sci. Technol.*, **B36**, 561–80.

Whittle, P. (1955) An investigation of periodic fluctuations in the New Zealand rabbit population. *N. Z. J. Sci. Technol.*, **B37**, 179–200.

Eighteen

Badger damage: fact or fiction?

R.G. SYMES

SUMMARY

Views on the significance of problems caused by badgers have varied widely. This chapter reports on some 1500 records for 1985–6 received by the Ministry of Agriculture's Wildlife and Storage Biology Discipline as requests for advice or complaints of damage alleged to have been caused by badgers throughout England and Wales. Analysis of these records has shown that more advice is given on badger problems in some parts of the country than on any other mammal or bird species. Four main categories of damage are recognized: direct structural damage, harm to crops, predation and behavioural problems. The significance of each of these is discussed and the status of the badger as a pest species is re-evaluated.

INTRODUCTION

No other British wild mammal has attracted so much attention as has the badger (*Meles meles* (L.)) in the last 20 years. During this time the species has gone from having little legal protection, first to being given limited protection under the Badgers Act 1973, then to achieving special protection, but in one county only, in 1979. Subsequently there were proposals to introduce further areas of special protection but the Wildlife and Countryside Act 1981 altered the badger's status completely by affording it substantial protection throughout the whole of England and Wales. In terms of rights of occupiers or others to take action against badgers, the situation changed from that under the original Badgers Act 1973 where an 'authorized person' was, within certain restrictions, entitled to kill badgers and was not required to give a reason for so doing, to that following the Wildlife and Countryside Act 1981 where 'taking or killing' of badgers now requires a licence unless the action is taken as an immediate, unpremeditated, defensive response.

This substantial increase in legal protection followed that of public interest in the species: a popular image was widely promoted of a beleaguered animal, seen by some to be a species close to extinction. Publicity given to cases of illegal badger baiting and badger digging attracted sympathy, as did the fate of the animal in areas where the Ministry of Agriculture, Fisheries and Food (MAFF) was carrying out its control operations against badgers, which were believed to be the source of bovine tuberculosis infection in cattle (Zuckerman, 1980; Anderson and Trewhella, 1985; MAFF, 1986; Dunnet, Jones and McInerney, 1986).

However there is, and always has been, a body of opinion which claims that some badgers cause unacceptable damage. Neal (1986) and many earlier writers reported instances where badger behaviour caused problems, but usually they attributed such action to individual 'rogue' badgers. Some farmers, and residents of small towns and villages in South West England, especially in Devon and Somerset, have been particularly forthright in condemning badgers for damage caused to their particular interests, and frequently such views have been linked with opinions that badger numbers have increased in recent years. One local authority in Somerset was reported to be ready to seek an amendment to the legislation to allow more action to be taken against badgers, and in 1986 the National Farmers Union (NFU) in Devon called on the Ministry of Agriculture, Fisheries and Food to make control licences more readily available, because they saw an escalation in the badger population with associated increases in damage. The allegations of damage referred especially to tunnelling action, predation on livestock, and feeding and territorial behavioural activities.

Estimates of the badger population of Britain are notoriously variable. Some have suggested that since about 400–1000 dead badgers (the result of traffic accidents), have been collected each year since 1975 by MAFF staff from roads of South Western England to be examined for the presence of bovine tuberculosis, that this indicates a substantial population in that region at least. Others by contrast have voiced concern about declining numbers, even in the South West. Neal (1986) published a map of densities of badger setts whose locations had been recorded throughout England, Wales and Scotland. Whilst he stressed that estimates of population density should only be based on numbers of social groups present, not on numbers of setts recorded, the map is generally accepted as an indication of probable relative badger numbers. The south and west of England and Wales are particularly represented in the 'common' and 'very common' categories, whereas in some counties to the east and north, badgers are considered to be 'infrequent', 'scarce', or 'absent'. These categories are not quantified but Neal reported that in exceptionally favourable districts, badgers sett densities may exceed 100 per 10 km^2 and that in more typical areas which may be described as good badger country 50 setts per 10 km^2 may be found. Unfortunately there is no reliable base line as yet on which comparisons may be made of changes in population densities.

It is against this background that Wildlife and Storage Biologists (WSB) of the Agricultural Development and Advisory Service (ADAS) of MAFF give advice on badgers and on problems attributed to them. This chapter focuses particularly on advisory requests during the years 1985 and 1986 from throughout England and Wales, with in addition some limited information from 1984 and 1987, and also reports the preliminary results of a survey of farmers in 1987. It discusses the nature of damage reported to have been caused by badgers, and the action taken by individuals to attempt to prevent or alleviate such damage.

SOURCES OF DATA

Since 1985, ADAS has maintained a national computerized database COSTER on which key information on advisory requests and complaints about problems with mammals, birds, and storage insects and mites is recorded. A precursor of COSTER, set up in 1984 in the south west only, provides some limited further information for that year also. The COSTER system is still being modified, but the data available include date, map reference, county, enquirer type (e.g. farmer), species, commodity affected or nature of enquiry. Each COSTER entry has a specific reference number which permits access to a more detailed report on the original field record sheet.

When enquirers, whether farmers, private householders or organizations, consider that damage caused by badgers is particularly serious, or could become so, they may formally apply to MAFF for a licence to 'kill or take' badgers. WSB staff advise their executive colleagues on the technical issues involved, since the latter are responsible for issuing such licences, and consult local officers of the Nature Conservancy Council (NCC). Such licence applications provide a secondary source of information about perceived problems with badgers. The outcome of licence applications in the period 1982 to 1987 is also discussed here.

Finally, the NFU in Devon discussed with MAFF the concern of some of its members about damage attributed to badgers, and about apparent increases in numbers of badgers in recent years. ADAS assisted the NFU in designing a questionnaire which the NFU sent to its members in Devon in 1987 to seek their views. I am grateful to the NFU for allowing me to report some of their preliminary results from the survey.

RESULTS

Incidence of badger damage

A total of 2071 advisory enquiries regarding badgers were made to WSB in England and Wales during the period 1985–87. In the two years 1985–86 for which most data are currently available, 1073 enquiries (73% of the total for those years), came from the South of England, with 714 (48%) being from South Western counties. Farmers (40%) and householders (43%) were the main groups seeking advice from WSB in 1985–86. A total of 553 advisory visits were made nationally in those two years. Data for 1984 for the South West only recorded 291 advisory requests on badgers. None of these figures included contacts or visits associated with MAFF's operational work connected with badgers and bovine tuberculosis in cattle. However, the number of reports received does perhaps reflect increased awareness of badgers as a result of concern expressed about the disease in non-operational areas.

The 1987 NFU survey sought the opinions of their members in Devon of the status of badgers on their farms. Of the 1030 respondents 83% said that badgers lived on or used their farms, 13% said there were none, and 4% did not know. Whilst 62% of those with badgers believed that numbers had increased over the last 5 years and 34% thought there had been no change, only 3% considered there had been a decrease in numbers. Although there are obvious questions to be posed on bias in the sample, it is nevertheless clear that some 500 farmers throughout Devon believe that badger numbers have increased over the last 5 years. The NFU survey asked if members had experienced nuisance or problems caused by badgers. Of those with badgers on their land 75% had experienced problems, and 30% perceived these as 'severe'.

Nature of badger damage

In the NFU's survey structural damage (42%) was the most frequent complaint, closely followed by crop damage (39%) and by predation (14%) (Figure 18.1). ADAS data showed that in 1985–86 enquiries were dominated by behavioural problems (25%), followed by feeding damage (18%), predation (13%) and structural damage (11%). Concern about disease involved 17% of enquiries in this period (Figure 18.2). The 595 COSTER entries for 1987, analysed differently, show behavioural problems and feeding damage in garden situations represented 30% of enquiries, damage to forestry, agricultural and horticultural crops represented 21%, structural damage 10%, and problems relating to livestock 7%. Concern about bovine tuberculosis dropped to 4% in 1987 (Figure 18.3). The remaining enquiries in 1987 mainly related to control techniques and legislation and cannot be attributed specifically to the categories described.

In garden situations the most common causes of concern are digging of

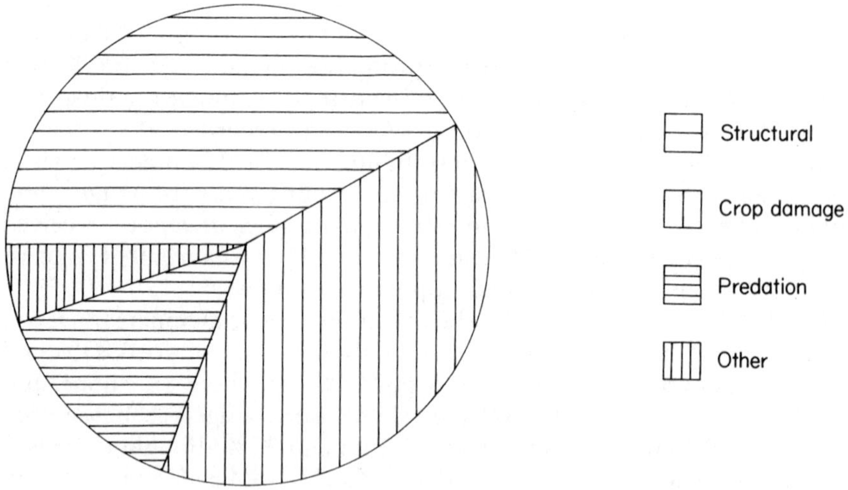

Figure 18.1 Types of damage recorded in 1987 Devon NFU badger survey (n = 1251).

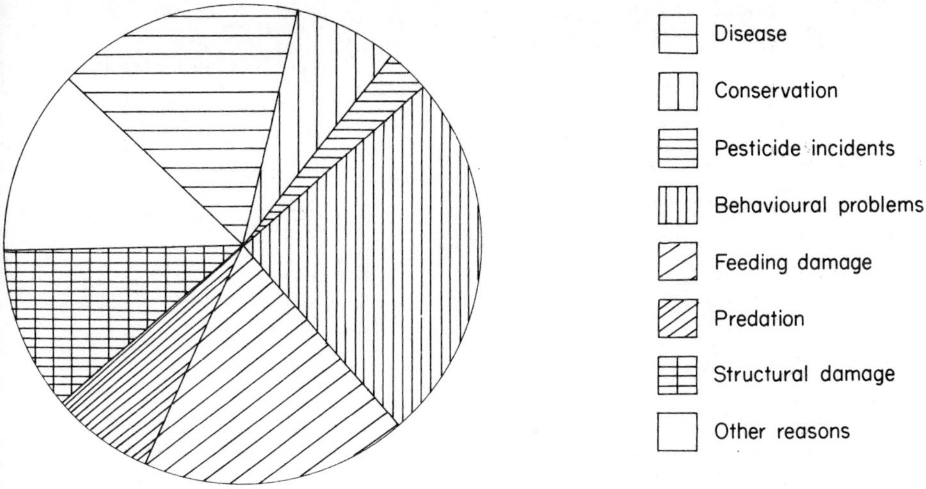

Figure 18.2 Reasons for requests for advice on badger problems made to ADAS in 1985–86 (n = 1073).

faecal pits, scratching up of lawns, tunnelling under fences, and feeding on produce such as strawberries and carrots. The deposition of faeces and scratching may in some cases by associated with territorial boundary marking and the areas affected can be substantial in relation to the size of the gardens. Similarly, indirect damage to pasture and golf courses by badgers scratching for invertebrates can be spectacular. Milner (1967) estimated that damage to one pasture amounted to at least a 10% loss of grass.

Farmers and official organizations are particularly concerned about tunnelling activities, both the tunnels and the resulting spoil heaps producing problems. A large sett was excavated by MAFF and the volume of the tunnel system was found to be 15.28 m^3, equivalent to 25 tonnes of soil excavated (MAFF, 1979). The depth of a 30–40 cm diameter tunnel below the ground surface is most important: shallow systems generally result in greater problems than deeper tunnels because of the lack of structural support by the roof. Tunnels under houses, and other buildings, in flood banks, railway embankments and under major roads cause particular concern. A road in Avon, undermined by badgers, cost £25,000 to re-construct, with an artificial badger tunnel. Farmers additionally face problems with livestock and farm machinery falling into shallow tunnels: in the former category valuable stock have suffered broken legs, whereas in the latter broken axles have occurred and there is also a risk to human life, to the extent that some farmers have been unable

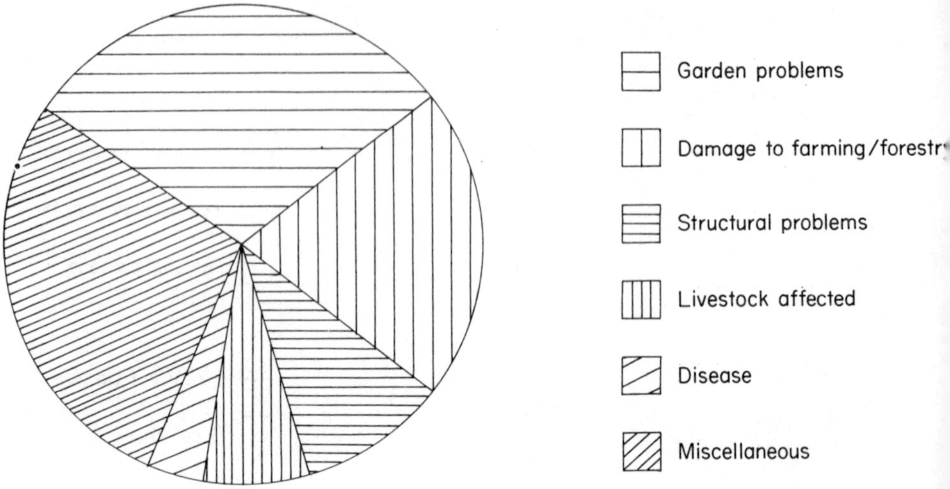

Figure 18.3 Reasons for requests for advice on badger problems made to ADAS in 1987 (n = 595).

to obtain insurance cover for working in fields known to contain badger tunnel systems. Spoil heaps frequently block drainage channels.

Damage caused by feeding on high-value horticultural crops such as strawberries or grape vines can be significant (e.g. Chapter 19), crops being attacked close to the harvest stage. Feeding on cereals such as oats may lead to large areas being flattened and beyond even the modern combine harvester to collect.

Most reports of predation involve poultry, but domestic pets such as guinea pigs are also taken. Lamb predation is frequently alleged, but post-mortem examination is needed in each case, to establish whether an animal was killed or scavenged by badgers, and fresh carcases have rarely been made available for examination.

Licences to prevent badger damage

Occupiers of land in England and Wales may apply to MAFF for licences to kill or take badgers for the purposes of preventing serious damage to property, crops, livestock etc. The legislation does not specify what constitutes serious damage. Since the enactment of the Wildlife and Countryside Act 1981 licences have been issued, following consultation with the NCC, to resolve 27 cases involving badger damage, and a further 23 formal licence applications have been refused (Table 18.1). Whilst

Table 18.1 Badger damage licence applications in England 1982 – November 1987

Year	Cases licenced	Cases where badgers killed	No. of badgers killed	Translocation cases	Licences refused
1982	3	1	7	1	—
1983	11	3	4	3	5
1984	2	0	0	1	6
1985	4	1	2	0	8
1986	6	0	0	4	0
1987	1	0	0	0	4

badgers were translocated to new areas in nine cases, they were killed in five cases only, no killing or taking occurring in the other cases.

Damage prevention

The COSTER data system has not recorded the techniques used to prevent damage by badgers. Information presented here has been compiled from a selection of original record sheets and from discussions and cannot be readily quantified.

Proofing measures at pet enclosures and small poultry units has been successful in reducing predation, but when the area to be protected has increased the cost has escalated considerably. Permanent fencing should be partly buried but may still be breached by badgers. Access to food sources has been denied by fencing or proofing techniques, or in some cases worm killers or insecticides applied to grassland or lawns may have reduced damage by badgers by reducing food supplies. Chemical repellents have been advocated by many and used widely with varying degrees of success. However, the substances which have been used in the past are not approved pesticides for repelling badgers under the Food and Environment Protection Act 1985, and from 1 January 1988 their use has been illegal. Repellents which are approved for use against some mammals include products with the following active ingredients : Ziram, Quassia, aluminium ammonium sulphate, bone oil and pepper (MAFF/HSE, 1986). Products which have been approved for use against badgers include: Narsty, Curb, Guardsman L Crop Spray, Liquid Curb Crop Spray, Guardsman STP, Stay Off (all based on aluminium ammonium sulphate) and Rabbit Smear Liquid (based on bone oil). However, their efficacy against badgers is largely untested. The use of repellents at setts, and the catching and translocation of badgers to new areas, all raise important humaneness issues which themselves can cause emotions to rise as much as do the problems they are invoked to resolve. The use of electrified fencing has usually been successful and some householders or groups of householders have been prepared to spend £70 to £100 to defend their gardens against badgers, but an estimate of £850 for a farmer to protect his oat fields effectively was not acceptable. The issues involved in preventing damage by badgers and humane methods of control were discussed by Symes (1985).

DISCUSSION

Whilst some problems faced by farmers, householders and others would appear primarily of nuisance value rather than of economic significance, others are undoubtedly more important and in some cases serious. Harris (1984) listed nuisances experienced by householders in Bristol, ranging from garden produce eaten, holes dug, pets attacked and fences broken. Macdonald (1984) found that the most common causes of complaints from his Midlands farm sample were trampling and rolling of crops, and tunnels collapsing under vehicles. However, whereas over half of Macdonald's Midlands farmers welcomed the presence of badgers on their land, only a quarter of the NFU's Devon sample of farmers described them as welcome. Many householders welcome badgers to their gardens and put out food for them. There have been demands for housing developments to be built around badger setts rather than disturb the badgers. Some new residents will no doubt welcome this, others will find difficulty in living with badgers. Step (1936) said the badger was 'exceedingly clean in his personal habits, and to prevent defilement of his sett digs pits in the neighbourhood for offensive waste'. Unfortunately many of the badger's neighbours are likely to find the waste offensive also.

A decision on whether damage is 'serious' is needed before a licence to kill or take badgers can be issued under the Badgers Act 1973 as amended by the Wildlife and Countryside Act 1981. Whilst many would agree that the deposition of faeces on a lawn is trivial, to someone who has taken pride in tending that lawn for many hours, the event is serious. When a badger breaks into a poultry unit and kills chickens, real damage occurs, but many would argue that killing a badger in such circumstances is not justified and that proofing of the poultry unit is required. The alternative of trapping the 'rogue' alive and translocating to a new area introduces new questions, particularly of the humaneness of such action. There is no straightforward economic formula which will determine the losses to be suffered, or anticipated, before the situation justifies the killing of a protected species. Each case is inevitably different, and has to be determined on the available evidence. The fact that in the period 1984 to 1987, when it is known that more than 2362 requests for advice on badger problems were received by WSB, only 13 cases resulted in the issue of a licence to 'kill or take' reflects: (1) the sometimes transient nature of problems; (2) the success achieved with some non-killing methods; and (3) the real practical difficulties faced by individuals in taking action under licence when translocation or killing is authorized.

CONCLUSION

Whether badger damage is seen as fact or fiction depends on individual views and attitudes. Many reports to WSB reflect concern and frustration over badger activities. Many other people are well prepared to tolerate the presence of badgers, or welcome them. The issue remains one which has to be handled with sensitivity and understanding of various viewpoints. Many people rely to a surprising extent on Press reports for their views on the status of wildlife species. As the protection afforded to British wild mammals increases, so will the requirement for further collection of accurate information on wildlife, and on the complex problems involved in preventing damage and controlling them in a humane and acceptable way.

ACKNOWLEDGEMENTS

I am grateful to my colleagues in WSB for the information provided from advisory and statutory cases. Michael Harrison collated much of the data reviewed here. Dr R. Putman, F.A. Hunter and Dr J.M. Way made valuable comments on early drafts.

REFERENCES

Anderson, R.M. and Trewhella, W. (1985) Population dynamics of the badger (*Meles meles*) and the epidemiology of bovine tuberculosis (*Mycobacterium bovis*). *Phil. Trans. R. Soc.* **310**, 327–81.

Dunnet, G.M., Jones, D.M. and McInerney, J.P. (1986) *Badgers and Bovine Tuberculosis – Review of Policy*. HMSO, London.

Harris, S. (1984) Ecology of urban badgers *Meles meles*: Distribution in Britain and habitat selection, persecution, food and damage in the City of Bristol. *Biol. Cons.*, **28**, 349–75.

Macdonald, D.W. (1984) A questionnaire survey of farmers' opinions and actions towards wildlife on farmlands in *Agriculture and the Environment* (ed. D. Jenkins), ITE Symposium No. 13, Institute of Terrestrial Ecology, Cambridge, pp. 171–77.

MAFF (1979) *Bovine Tuberculosis in Badgers*. 3rd Report.

MAFF (1986) *Bovine Tuberculosis in Badgers.* 10th Report.
MAFF/HSE (1986) *Pesticides 1986.* Reference Book 500. HMSO, London.
Milner, C. (1967) Badger damage to upland pasture. *J. Zool., London,* **153**, 544–6.
Neal, E.G. (1986) *The Natural History of Badgers.* Croom Helm, London and Sydney.
Step, E. (1936) *Animal Life of the British Isles*, Warne, London.
Symes, R.G. (1985) Humane control of badgers in *Humane Control of Land Mammals and Birds.* UFAW, pp. 118–23.
Zuckerman, S. (1980). *Badgers, Cattle and Tuberculosis.* HMSO, London.

Nineteen

Badgers as pests in English vineyards

T.J. ROPER, P. LÜPS AND S. LYCETT

The pest status of badgers (*Meles meles*) was investigated during a five-year study at a vineyard in E. Sussex. Analysis of 480 faeces samples from 6 badger territories in the region of the vineyard revealed that grapes were consumed during September, October and November in each of 3 years. Overall, grapes were estimated to constitute 2% of the diet by volume. However, if the analysis was confined to faeces from territories shown by bait-marking to include parts of the vineyard, and to the period when grapes were available, grapes constituted 64% of the diet by volume. Thus grapes were not an important part of the diet year-round in the area as a whole, but they were eaten regularly and in large volumes when readily available. Observation of 3 radio-collared animals confirmed that badgers took grapes directly from the vines, spending from 30 min to 2 h in the vineyard each night. Many more grapes were knocked to the ground than were eaten and in addition badgers damaged nets used to exclude birds. During the last 2 years of the study it proved possible to prevent badger damage by surrounding the vines with a low electric fence (total length 1300 m) consisting of 2 strands of 3-filament 'Polywire' supported at heights of 12 cm and 22 cm respectively above ground level. The fence was powered by a mains supply of maximum output 1.3J.

In order to assess the importance of badger damage nationally a questionnaire was sent to 100 vineyards distributed throughout the south of England. Of 96 replies received, 22% reported some degree of badger damage. Damage was most frequent and most severe in the west and south-east of England where independent evidence suggests that badger population densities are high. Of those vineyard owners who reported that badgers were 'common' in their locality, 92% suffered damage. The most common causes of complaint against badgers were that they ate or otherwise damaged grapes (13% of replies) and/or damaged nets and fences (14% of replies). Estimates of the financial cost of badger damage ranged from £50 to £4000 pa but were mostly in the range £100 to £1000 pa. Five vineyard owners had suffered severe badger damage in the past but had taken measures to prevent it, either by means of electric fences or by growing their vines on high trellises. It appears that badgers can constitute a serious nuisance for vineyards in parts of the country where badgers are numerous.

Twenty

The control of rabies in urban fox populations

G.C. SMITH AND S. HARRIS

SUMMARY

At present the occurrence of foxes in urban environments is predominantly a British phenomenon, and fox populations in British urban areas have some of the highest densities recorded in any habitat in Europe or North America. This chapter briefly introduces the rabies virus and reviews past attempts at rabies control in Europe and North America. If the disease is reintroduced to Britain it will probably be via a smuggled pet, and the resulting epizooty might originate in an urban area. The effects of such a single point source of infection on the pattern of spread of rabies in Britain are discussed, as are possible methods of fox control in an urban environment. The Canadian 'Ontario Rabies Model', with parameters relevant to British urban situations, was used to evaluate some of the problems associated with rabies control in an urban area. The value of such theoretical modelling in planning methods of disease control is discussed.

INTRODUCTION

Rabies is a virus which can affect all warm-blooded animals. It is primarily a disease of the central nervous system and its characteristic symptoms result from the spread of damage in the brain and spinal cord (Brown and Crick, 1978). The occurrence of the symptoms of rabies in foxes, dogs and man almost invariably preludes death and very few individuals, if any recover. Steck and Wandeler (1980) report that less than 2% of experimentally inoculated foxes survived to become immune. Contrary to popular belief only about half of all foxes show the furious aggression normally associated with rabies and many other symptoms, including paralysis, fits, lack of fear, vomiting and trembling are possible (George *et al.*, 1980). In the fox, an animal particularly susceptible to the rabies virus, no symptoms appear for between 10 and 110 days, with an average of about 20 to 30 days; this is followed by a period of three to five days during which symptoms are present and the virus is excreted (Sikes, 1962; Tierkel, 1966; Parker and Wilsnack, 1966; Black and Lawson, 1970).

The present epizootics in both North America and Europe are sylvatic, and in any one area are primarily confined to only one or two species. In Europe it is the red fox (*Vulpes vulpes*) which is responsible for spreading and maintaining the disease, although rabies is also found in badgers, stone-martens, polecats, roe-deer and small rodents (Wandeler *et al.*, 1974). The current European epizootic originated in Poland about the time of the Second World War. It spread westward, along a broad front at up to 60 to 80 km per year, reaching Denmark in 1964, Belgium, Luxemburg and Austria in 1966, Switzerland in 1967 and France in 1968. Since continental countries have been forewarned of the approach of rabies they have had time to organize and implement costly control schemes.

Most of these countries have not been able to eliminate rabies, although there is some evidence that the control policies have reduced the incidence of rabies. For instance, in France the use of fox bounties, poison baits, gassing of earths and night shooting appears to have contributed to the relative stability of the rabies front in recent years (Roboly, 1982; Toma and Lignières, 1985). Denmark has successfully eliminated rabies from southern Jutland on three occasions by an intensive campaign of gassing, poisoning and shooting in a coast to coast zone up to 60 km in depth (Müller, 1971; Westergaard, 1982). Italy has also contained the spread of rabies on a number of occasions by reducing fox populations in the passes from Austria (Irsara *et al.*, 1982).

Rabies control campagins are not designed to kill all the foxes present in an area, but to reduce the population density to a level such that each rabid fox infects, on average, less than one susceptible fox, thus continually reducing the incidence of rabies until it dies out completely. This critical level is believed to be between 0.25 and 1 adult fox km^{-2} (Anderson *et al.*, 1981; Bögel *et al.*, 1981).

This reduction in the number of susceptible foxes can also be achieved by vaccination. Inactivated or attenuated rabies virus vaccines have been available for some years but their use has been restricted by reservations about their efficacy (Baer, Abelseth and Debbie, 1971; Wachendörfer, Kiefert and Frost, 1982; Frost, Friedrich and Wachendörfer, 1982; Pépin *et al.*, 1985). However, newer recombinant and live vaccines are more promising (Schneider, 1985; Blancou *et al.*, 1986). The use of vaccines has advantages over population reduction as a means of rabies control, since vaccination in response to a rabies event would not reduce the numbers of other species nor would it change the dispersal behaviour, age and sex ratios of a healthy fox population.

Any introduction of the disease into foxes in Britain is likely to originate from a smuggled pet, and any resulting sylvatic infection will probably produce a radial spread from a point source. The ensuing control operation will therefore be approximately circular centring on the location of the first report of rabies. The primary objective of any control campaign in Britain is to halt and eradicate the outbreak of rabies immediately upon its detection (Waterhouse, 1971). The experience in continental Europe has shown that the control of rabies is most successful when dealing with a point source of infection where rabies has spread along a narrow corridor such as an isthmus or mountain pass (Irsara *et al.*, 1982); less success has been achieved when rabies advances along a broad front since any control operations must be more extensive.

When dealing with a rabies infection from a point source, the radius of the control zone will depend on a number of factors: the time of year, fox population density, time elapsed before detection and the certainty of detecting the centre of the radial spread. Since the experience on the continent has shown that the likelihood of reporting a rabid fox is low, it may be up to five or six months before the authorities are aware of the situation. The longer the time between the introduction and detection of the disease, the greater the required area of the control operation. As an example of the potential rate of spread, Murray, Stanley and Brown (1986), using estimated fox densities, calculated that an uncontained rabies outbreak originating in Southampton would reach Manchester within four years.

Rabies control methods in the United Kingdom are likely to be similar to those used on the continent. Although the use of vaccines may be feasible in the near future, at present population reduction methods will be employed. However, fox control in urban areas poses particular problems. Coventional techniques are unlikely to achieve significant population control (Harris, 1985) and so poisoning, in association with other methods where appropriate, is likely to remain the favoured control option in urban areas for the immediate future. Since urban foxes are common in many British cities but much rarer in continental Europe (Harris, 1977), there is little previous experience about the effects of rabies on urban fox populations. In the absence of this direct experience, we have used a simulation model to study the potential pattern of rabies spread in urban areas, and to examine the levels of population control necessary to eliminate sylvatic rabies. The preliminary results are presented in this chapter.

METHODS

Although there are some advantages to using a model with very few parameters, such models are only valid if the biological assumptions are realistic. For instance such models may accurately simulate the rate of spread of rabies under constant conditions but may not produce realistic results when conditions such as population density and level of control of the fox population change. In general, when a population is subjected to artificially high mortality it would be expected that individual fecundity increases and the probability of mortality from natural causes decreases. A realistic/useful model would therefore have to make some adjustments to the relevant population parameters to simulate these circumstances. Also, the velocity of rabies spread presumably depends upon the density and composition of a fox population and their contact rates; but, during a period of intensive population control these factors will change, thereby affecting the rate of spread of rabies. An effective model will necessitate a spatial component to model fox dispersal and infective contacts, and will need to include a stochastic element so that the probability of any event may be calculated.

With these considerations it was decided to utilize a simulation model which could cause a variation in output depending upon the specific details of the fox behaviour and distribution. We required a model which could simulate an open population (one capable of immigration and emigration) of foxes over the wide variation in population densities such as those found in British cities; mean densities range from 0.1 to 2.8 fox family groups km^{-2} (Harris and Smith, 1987a). One such model has been developed, the Canadian 'Ontario Rabies Model' (see Voigt, Tinline and Broekhoven (1985) for a detailed description of this model, its construction and the assumptions made). We used this model to perform a preliminary analysis of the effects of rabies in a homogeneous urban fox population and to look at the effects of fox population control during a rabies event. In all cases any ingressing foxes were assumed to be healthy since we were examining the focal outbreak of a rabies incident. The population parameters used in this model are those from fox population studies in Bristol (Harris and Smith, 1987b; Harris and Trewhella, 1988; Trewhella, Harris and McAllister, 1988; Trewhella and Harris, 1988; Harris, unpublished data). Since one objective of this project was to attempt to calculate the required area and level of any fox control campaign we decided to run the model in a worst case simulation. Rabies was injected into the simulation by adding four juvenile rabid foxes at the beginning of autumn, immediately prior to dispersal, and discovery and thus any fox control delayed until winter, giving a time lag of three months. This time interval was chosen because Bacon (1981) suggested that there could be a delay of two to seven months before the disease is detected.

We decided to examine the effects of changing three important parameters of the model separately: cell size, contact rate and management by fox removal. The model operates on a grid where each cell is designed to represent a home range. Changing the size of each cell, therefore, is roughly equivalent to changing the size of the home ranges while maintaining a fixed density. This effectively varies the degree of home range overlap between family groups and thus the number of possible contacts a fox may have. However, due to the design of the model, this representation is only true if within-group and between-group contact rates are held at the same value. This has been done for all runs of the model at different cell sizes. Contact rate, measured here as the probability of an individual rabid fox infecting each neighbouring fox, is an unknown parameter in sylvatic rabies spread. Therefore the contact probability was adjusted from 0.00 up to 1.00 in various runs of the model for all the densities analysed. During spring within-group contact probability is set to 1.00 to simulate the high within-group contact rate associated with this period.

To investigate the effects of a control campaign, additional analyses were undertaken where some control of the fox population was attempted. Since these studies were worst case simulations, any management activity was delayed until early winter when a set proportion of all foxes were killed. A second removal campaign in early spring was instigated in half of the simulations, which were thus designed to examine the effects of campaigns with two control periods. The proportion of foxes removed in the first and second control campaigns was kept similar. All simulations were performed with an area equal to that of urban Bristol (10.5 km × 11.0 km) except for those which examined the effectiveness of increasing the area of fox control.

To express the results of each rabies incident it was found necessary to present them in three ways. First, the introduction of a rabid fox into the simulation will not lead to a rabies epidemic if, purely by chance, it fails to infect any other individual. This is presented as the percentage of cases which fail to lead to an epizootic. Once a rabies outbreak occurs it may be considered from two points of view: the effect on the human population, i.e. the duration of the rabies incident and therefore the length of any enforced control measures, or the effect on the fox population. The latter is presented as the proportion of the original population which becomes infected during the incident, i.e. summing the total numbers of infected foxes in each year and comparing them to the original fox population. In this study all three sets of results were considered to obtain a comprehensive picture of the outbreak.

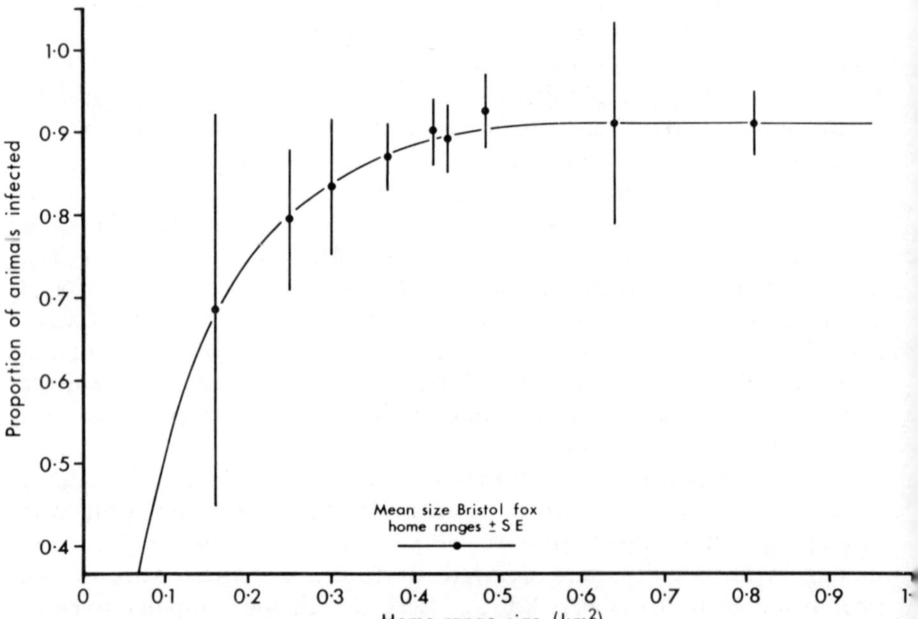

Figure 20.1 Graph of the sum of all infected foxes as a proportion of the original population plotted against home range size; the vertical bars represent the SD from at least 50 simulations. The horizontal bar represents the mean size ± SE of fox home ranges found in Bristol.

RESULTS

Effects of changing home range size

In Bristol home range sizes have been measured from 25.7 ha to 78.2 ha, with a mean of 45.4 ± 6.9 ha (Harris, 1980). In grid form this equates to each square having sides of 0.5 km to 0.88 km. At least 50 simulations were run at each of the following grid sizes: 0.40, 0.50, 0.55, 0.60, 0.65, 0.66, 0.74, 0.80 and 0.90 km, with a mean fox density equal to that found in Bristol. The proportion of infected animals is shown in Figure 20.1 and the duration of the rabies event in Figure 20.2. All these simulations were run with a probability of a rabid fox infecting other foxes, both within and between cells, of 0.50. Both graphs show that rabies is not more pronounced when home range sizes are greater than the mean, but reduction of home range size, which in the model reduces the number of within and between group contacts and leaves some areas unoccupied,

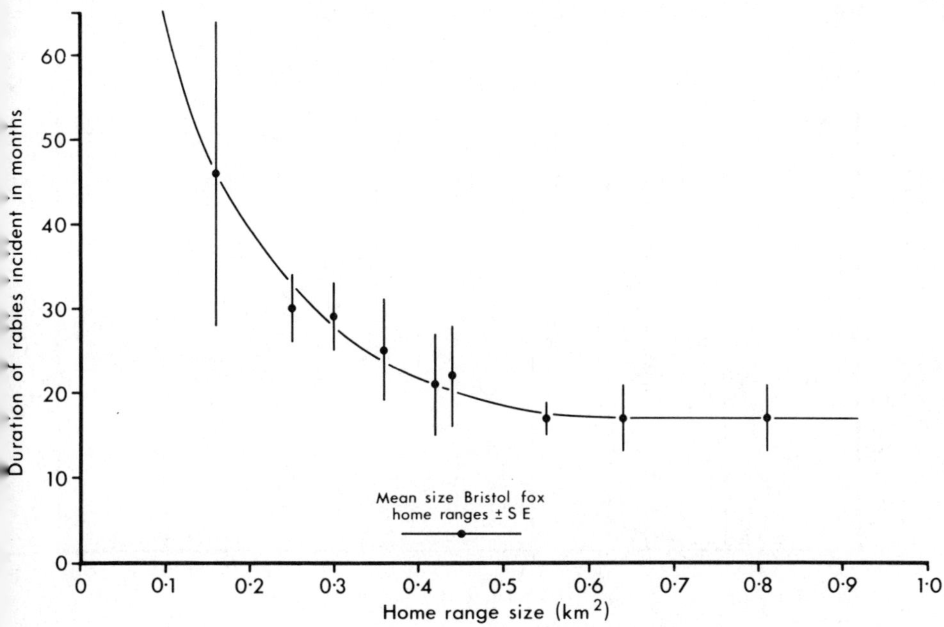

Figure 20.2 Graph of the duration of the rabies incident in the simulated population (in months) plotted against home range size; the vertical bars denote the SD from at least 50 simulations. The horizontal bar represents the mean size ± SE of fox home ranges found in Bristol.

progressively leads to a reduction in the overall effect of rabies on the fox population. Since the effect on the fox population reduces with decreasing home range size, the duration of the rabies event is lengthened. With all values of home range size equal to or greater than the mean value for Bristol, the fox population is reduced to between 1 and 2% of its original size within one and a half years of the outbreak.

Effects of changing contact probabilities

A minimum of 50 simulations were run each time the probability was adjusted. Two fox densities were used: 1.82 fox family groups km^{-2} and 0.91 fox family groups km^{-2}. The former is similar to that found in cities such as Bristol and Poole and the latter similar to that in cities such as Bath and Nottingham (Harris and Smith, 1987a). Figure 20.3 shows the percentage of simulations in which no new foxes were infected, and the proportion of infected foxes during epizootics for both densities. The mean duration of the rabies events for both densities is shown in

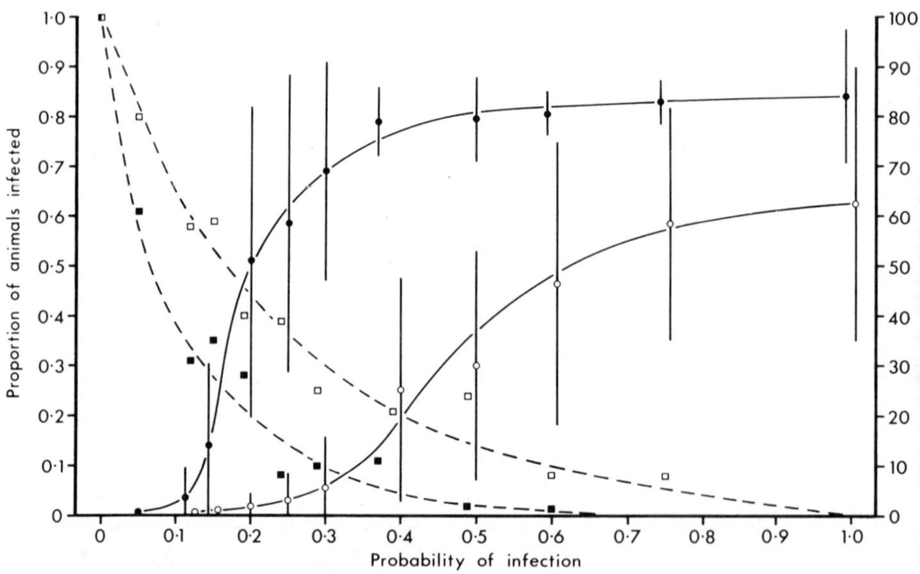

Figure 20.3 Graph of the sum of all infected foxes as a proportion of the original population (circles and solid lines) and the percentage of simulations where no healthy foxes were infected (squares and broken lines), against the probability of a rabid fox infecting a healthy fox. The closed symbols represent a fox density equivalent to that in Bristol (1.82 fox family groups km^{-2}) and the open symbols a density of 0.91 fox family groups km^{-2}. The vertical bars denote the SD from at least 50 simulations.

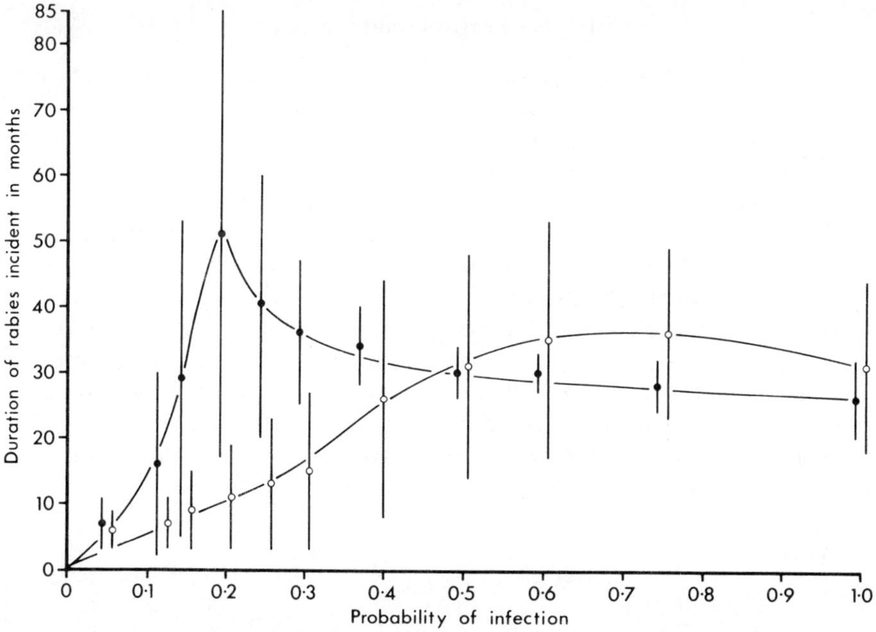

Figure 20.4 Graph of the duration of the rabies incident in the simulated population (in months), against the probability of a rabid fox infecting a healthy fox. The closed symbols represent a density of 1.82 fox family groups km^{-2} and the open symbols a density of 0.91 fox family groups km^{-2}. The vertical bars denote the SD from at least 50 simulations.

Figure 20.4. As one would expect, increasing the probability of infection increases the effect on the fox population, but an asymptote is reached where a further increase in the probability of infection has very little effect on the proportion of foxes that become infected during the epizootic.

The duration of the rabies incident within the population is very short if there is a low probability of infection, since only a few animals become infected at any time and chance fluctuations tend to lead to early and spontaneous eradication of rabies. A probability of near unity also leads to a short duration incident, since rabies has removed nearly all the foxes leaving a population only some 4% of the original size. Between these two values is a peak where rabies infects new individuals at a rate high enough to avoid early fluctuations to zero and low enough to avoid removing all the foxes. In these cases rabies may last for between one and a half and seven years and leave a population some 60% of its original size. In the lower density population this peak is at a higher probability of infection, at which point the rabies incident may last longer than in the denser population (Figure 20.4).

Effects of rabies management

The simulations used to examine the effects of fox control were identical to those above, but in addition two levels of fox control were added. In the first instance 50% of all foxes, healthy and infected, were removed at the beginning of winter, three months after the first foxes were infected. In the second case a further 50% of those remaining were removed in early spring, thereby removing 75% of the foxes over a six-month period. The simulations were again run at different probabilities of infection between zero and one, and the results (proportion of infected animals and duration of the incident) are given in Figures 20.5 and 20.6. Up to 140 simulations were performed at each level of fox control. The percentage chance of failing to produce an epizootic is exactly the same as for the simulations above since it cannot be affected by the control measures.

Although removing 50% of the foxes appears to reduce the proportion

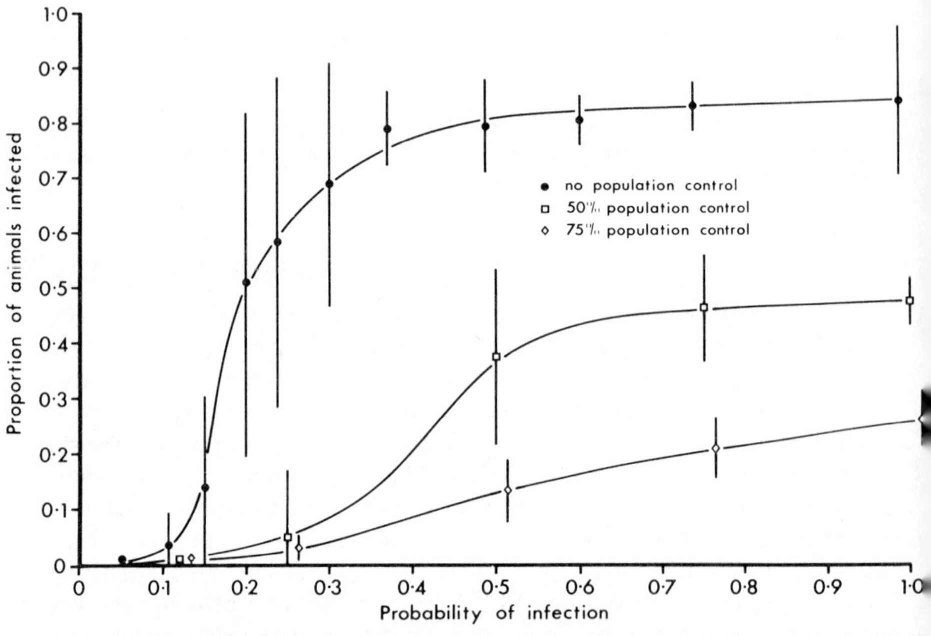

Figure 20.5 Graph of the sum of all infected foxes as a proportion of the original population plotted against the probability of a rabid fox infecting a healthy fox. The closed circles denote simulations with no population control, the open squares simulations with 50% population control three months after the onset of the epizootic, and the open diamonds simulations with 50% population control three months after the onset of the epizootic, and a further 50% (75% in all) after a further three months. The vertical bars denote the SD from at least 50 simulations.

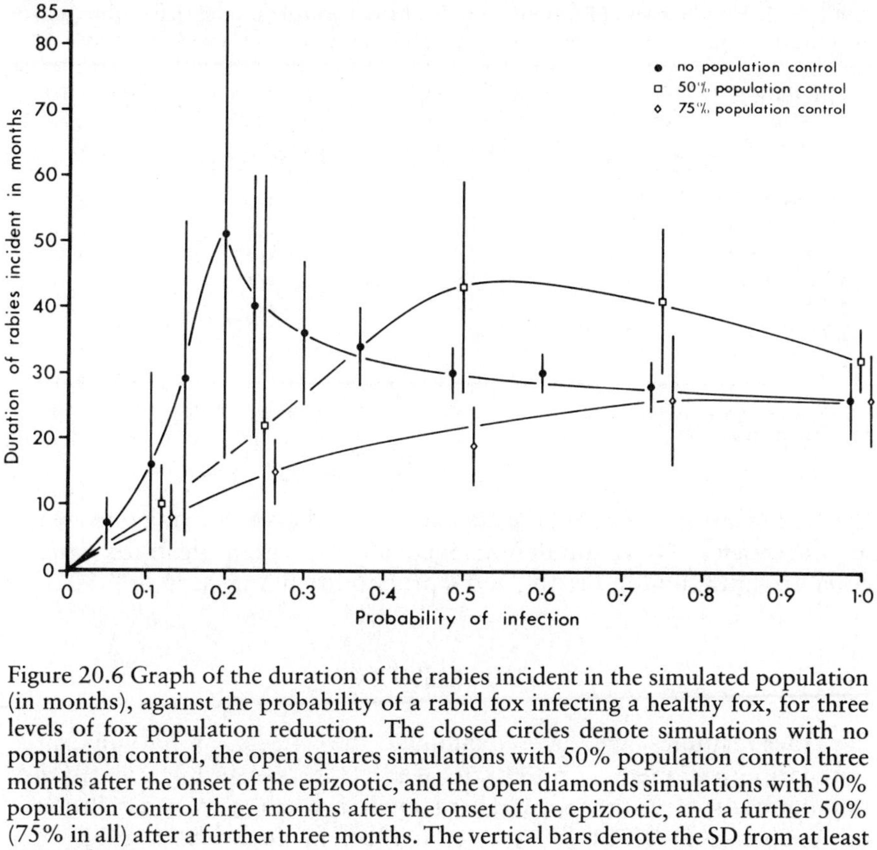

Figure 20.6 Graph of the duration of the rabies incident in the simulated population (in months), against the probability of a rabid fox infecting a healthy fox, for three levels of fox population reduction. The closed circles denote simulations with no population control, the open squares simulations with 50% population control three months after the onset of the epizootic, and the open diamonds simulations with 50% population control three months after the onset of the epizootic, and a further 50% (75% in all) after a further three months. The vertical bars denote the SD from at least 50 simulations.

of infected foxes by almost 50%, the duration of the incident is not consistently less until 75% of the foxes are removed. The reduction in the proportion of infected foxes after fox control is not merely an artifact of having fewer foxes present.

Since the fox population is not isolated, and is therefore open to emigration, the number of rabid foxes dispersing from the simulation area was recorded so that the chances of the infection spreading beyond the area of control could be assessed. The results are shown in Table 20.1, which shows the area of the simulation, the number of simulations run, the percentage of cases where no rabid foxes disperse outside the simulation area, the mean number of rabid foxes which do disperse outside the simulation area if the control is unsuccessful and the level of fox control applied to the entire area. As the level of control increases and

Table 20.1 Effectiveness of different levels of fox control and the size of the control area in containing a rabies event. For a list of the parameters used see the text

Simulation size (km)	Number of runs	% cases with no rabid emigrants	Mean number of rabid emigrants*	Fox control level (%)
10.5 × 11.0	40	3	11.6	0
10.5 × 11.0	37	11	3.8	50
10.5 × 11.0	59	47	1.6	75
15.0 × 15.0	47	66	1.3	75
17.5 × 17.5	59	83	1.1	75
20.0 × 20.0	65	83	1.0	75
22.5 × 22.5	56	95	1.0	75

* Calculated only for those simulations where at least one rabid fox dispersed outside the simulation area.

the area of control enlarges the chances of rabid foxes moving outside the area decreases. These simulations were all run with a mean fox density equal to that found in Bristol, and a probability of infection of 0.50.

DISCUSSION

For these studies we have examined the effects of varying three population parameters on the rate of rabies spread in high-density fox populations such as those found in many British urban areas. To illustrate the problems involved with rabies control in high-density urban fox populations, we have used worst-case simulations throughout the study. However, it is important to note that since we have no experience of the pattern of rabies spread in such high-density fox populations we have no criteria against which to evaluate our results.

Increasing the model's cell size has little effect on the spread of rabies throughout the population. Increasing home range overlap in a population effectively increases the available number of contacts per rabid fox, which is essentially the same as increasing the probability of infection. Figure 20.3 shows that this will not increase the proportion of infected foxes in a population with a density as high as that found in Bristol, if the probability of infection between foxes is moderately high. If the fox density was lower, then an increase in home range overlap would be expected to produce an increase in the proportion of infected foxes if density and probability of infection combined are below the plateau stage of the graph. If home range overlap is so large that a fox has a chance of meeting foxes in groups other than its immediate neighbours, then although the final number of foxes infected may not increase, the rate of spread of the infection would be expected to be greater, thus leading to a shorter duration rabies event. This implies that in lower density urban areas the probability of infection will effectively be increased if there are overlapping home ranges.

There are no quantified data on the contact rates of wild foxes, so it is necessary to examine a complete range of probabilities of infection. In high density areas such as Bristol (1.82 fox family groups km^{-2}), it appears that any calculated contact probability only has to be accurate if it lies between 0.1 and 0.4, since at rates lower than this the rabies event is short lived and affects few foxes, while at greater levels an increase in the probability of infection produces very little change in either the proportion of the population infected or the mean duration. For the lower density fox population (0.91 fox family groups km^{-2}) changing contact probabilities between 0.25 and 0.80 produces large changes in the rabies outbreak. Once some approximation of sylvatic contact rates has been calculated it will, therefore, be possible to determine for which fox densities, and thus for which cities, such an approximation will suffice.

To explain the existence of the plateau stage in Figure 20.3 it is necessary to look at the probability of a healthy fox becoming infected. If there are only two foxes, one rabid and one healthy, and the rabid fox infects healthy foxes with a probability of 0.5, the probability of the healthy fox becoming rabid is 0.5. This is effectively the case at the beginning of a rabies outbreak when there are only a few rabid foxes. As the epizootic continues and the number of rabid foxes increases, each healthy fox will have a chance to contract rabies from more than one rabid fox. In the above example, if there are two rabid foxes and one healthy fox, the healthy fox has a probability of becoming infected of 0.75 since it is exposed to two infected foxes. The plateau occurs once each healthy fox has a probability of becoming infected which is near to unity because of the number of possible rabid contacts it may have.

Any reduction in the number of foxes present will reduce the total proportion infected (Figure 20.5) but the duration of the rabies epizootic may be increased if not enough foxes are removed (Figure 20.6). From a control point of view the duration of the outbreak is also important. In these simulations, run at a fox population density of 1.82 family groups km^{-2}, at least 75% of the foxes need to be removed for the duration of the incident to decrease. Any further increase in the proportion of foxes removed will further decrease the duration of the epizootic. In lower density areas the minimum proportion of foxes that need to be removed to achieve the same effect will be lower.

The area of the control operation is also important when planning to control a rabies outbreak. In our simulations, we have only considered the effect on an isolated population. Table 20.1 suggests that for a homogeneous fox population of density 1.82 family groups km^{-2} and a probability of infection of 0.50, a control zone in which 75% of the foxes are killed, with an area of 506 sq. km, has a 95% chance of being totally successful and containing the epizooty. Such an approach to rabies modelling could be of particular value when trying to assess the size for a control zone.

The results of these simulation studies show that under certain circumstances a low level of fox control during a rabies outbreak may actually increase the duration of the outbreak, as well as run a high risk of permitting rabid foxes to disperse beyond the control area. They also demonstrate that the precise biological data on the exact probabilities of foxes infecting each other, which are currently unavailable, are only important in areas with lower fox population densities, since at high densities the probability of infection has little impact on the duration or effect of a rabies epizooty.

Our attempts to model the possible consequences of a rabies outbreak in urban foxes in Britain suffers from certain necessary drawbacks. In particular, the model we used could only realistically simulate a homogeneous fox density. Urban fox densities vary considerably from one area of a city to another (Harris and Rayner, 1986a), and the effect on the spread of rabies in such an environment is unknown. Future work will investigate this more complex situation. It is important that a model is developed whereby the probability of infection can be varied in some manner with the density of the fox population. Harris and Rayner (1986b) have shown how urban fox numbers can be predicted from simple habitat variables, and in a heterogeneous simulation model it will be possible to mimic rabies events in specified urban areas. It will then be possible to calculate the size of effective control zones for individual specific cities or set of circumstances, and to evaluate accurately the effectiveness of particular control strategies.

ACKNOWLEDGEMENTS

S. Harris is grateful to the Ministry of Agriculture, Fisheries and Food, the Natural Environment Research Council and the Nature Conservancy Council for funding his studies and G.C. Smith is grateful to the Science and Engineering Research Council for a studentship. Both authors are grateful to the many people who have assisted with the urban fox studies and to P.J. Bacon, W.J. Cresswell, W. Trewhella and D.R. Voigt for their comments on drafts of this manuscript.

REFERENCES

Anderson, R.M., Jackson, H.C., May, R.M. and Smith, A.D.M. (1981) Population dynamics of fox rabies in Europe. *Nature, Lond.*, **289**, 765–71.

Bacon, P.J. (1981) The consequences of unreported fox rabies. *J. Environ. Mgmt*, **13**, 195–200.

Baer, G.M., Abelseth, M.K. and Debbie, J.G. (1971) Oral vaccination of foxes against rabies. *Am. J. Epidemiol.*, **93**, 487–90.

Black, J.G. and Lawson, K.F. (1970) Sylvatic rabies studies in the silver fox (*Vulpes vulpes*). Susceptibility and immune response. *Can. J. Comp. Med.*, **34**, 309–11.

Blancou, J., Kieny, M.P., Lathe, R., Lecocq, J.P., Pastoret, P.P., Soulebot, J.P. and Desmettre, P. (1986) Oral vaccination of the fox against rabies using a live recombinant vaccinia virus. *Nature, Lond.*, **322**, 373–5.

Bögel, K., Moegle, H., Steck, F., Krocza, W. and Andral, L. (1981) Assessment of fox control in areas of wildlife rabies. *Bull. Wld Hlth Org.*, **59**, 269–79.

Brown, F. and Crick, J. (1978) Rabies and rabies virus. *Biologist, Inst. Biol.*, **25**, 91–7.

Frost, J.W., Friedrich, H. and Wachendörfer, G. (1982) Effectivity of oral vaccination of foxes against rabies with Flury HEP strain 675. *Comp. Immunol. Microbiol. Infect. Dis.*, **5**, 181–4.

George, J.P., George, J., Blancou, J. and Aubert, M.F.A. (1980) Description clinique de la rage du renard. *Revue Méd. Vét.*, **131**, 153–60.

Harris, S. (1977) Distribution, habitat utilization and age structure of a suburban fox (*Vulpes vulpes*) population. *Mammal Rev.*, **7**, 25–39.

Harris, S. (1980) Home ranges and patterns of distribution of foxes (*Vulpes vulpes*) in an urban area, as revealed by radio tracking in *A Handbook on Biotelemetry and Radio Tracking* (eds C.J. Amlaner and D.W. Macdonald), Pergamon Press, Oxford, pp. 685–90.

Harris, S. (1985) Pest control in urban areas: humane control of foxes in *Humane Control of Land Mammals and Birds* (ed. D.P. Britt), Universities Federation for Animal Welfare, Potters Bar, pp. 63–74.

Harris, S. and Rayner, J.M.V. (1986a) Urban fox (*Vulpes vulpes*) population estimates and habitat requirements in several British cities. *J. Anim. Ecol.*, **55**, 575–91.

Harris, S. and Rayner, J.M.V. (1986b) Models for predicting urban fox (*Vulpes vulpes*) numbers in British cities and their application for rabies control. *J. Anim. Ecol.*, **55**, 593–603.

Harris, S. and Smith, G.C. (1987a) The use of sociological data to explain the distribution and numbers of urban foxes (*Vulpes vulpes*) in England and Wales. *Symp. Zool. Soc. Lond.*, **58**, 313–28.

Harris, S. and Smith, G.C. (1987b) Demography of two urban fox (*Vulpes vulpes*) populations. *J. Appl. Ecol.*, **24**, 75–86.

Harris, S. and Trewhella, W.J. (1988) An analysis of some of the factors affecting dispersal in an urban fox (*Vulpes vulpes*) population. *J. Appl. Ecol.*, **25**, 409–22.

Irsara, A., Ruatti, A., Gagliardi, G., Orfei, Z., Bellani, L. and Mantovani, A. (1982) Le contrôle de la rage selvatique dans le nord-est de l'Italie. *Comp. Immunol. Microbiol. Infect. Dis.*, **5**, 327–35.

Müller, J. (1971) The effect of fox reduction on the occurrence of rabies. Observations from two outbreaks of rabies in Denmark. *Bull. Off. Int. Épizoot.*, **75**, 763–76.

Murray, J.D., Stanley, E.A. and Brown, D.L. (1986) On the spatial spread of rabies among foxes. *Proc. R. Soc. Lond. B*, **229**, 111–50.

Parker, R.L. and Wilsnack, R.E. (1966) Pathogenesis of skunk rabies virus: quantitation in skunks and foxes. *Am. J. Vet. Res.*, **27**, 33–8.

Pépin, M., Blancou, J., Aubert, M.F.A., Barrat, J., Coulon, P. and Flamand, A. (1985) Oral immunization against rabies with an avirulent mutant of the CVS strain: evaluation of its efficacy in fox (*Vulpes vulpes*) and its infectivity in seven other species. *Ann. Inst. Pasteur Virol.*, **136 E**, 65–73.

Roboly, O. (1982) Application du tir de nuit des renards à la prophylaxie de la rage. *Comp. Immunol. Microbiol. Infect. Dis.*, **5**, 389–95.

Schneider, L.G. (1985) Oral immunization of wildlife against rabies. *Ann. Inst. Pasteur Virol.*, **136 E**, 469–73.

Sikes, R.K. (1962) Pathogenesis of rabies in wildlife. I. Comparative effect of varying doses of rabies virus inoculated into foxes and skunks. *Am. J. Vet. Res.*, **23**, 1041–7.

Steck, F. and Wandeler, A. (1980) The epidemiology of fox rabies in Europe. *Epidemiol. Rev.*, **2**, 71–96.

Tierkel, E. (1966) Rabies in wild animals. *Symp. Ser. Immunobiol. Stand.*, **1**, 245–50.

Toma, B. and Lignières, R. (1985) Le programme français de lutte contre la rage. *Ann. Inst. Pasteur Virol.*, **136 E**, 483–90.

Trewhella, W.J. and Harris, S. (1988) A simulation model of the pattern of dispersal in urban fox (*Vulpes vulpes*) populations and its application for rabies control. *J. Appl. Ecol.*, **25**, 435–50.

Trewhella, W.J., Harris, S. and McAllister, F.E. (1988) Dispersal distance, home range size and population density in the red fox (*Vulpes vulpes*) – a quantitative analysis. *J. Appl. Ecol.*, **25**, 423–34.

Voigt, D.R., Tinline, R.R. and Broekhoven, L.H. (1985) A spatial simulation model for rabies control in *Population Dynamics of Rabies in Wildlife* (ed. P.J. Bacon) Academic Press, London, pp. 311–49.

Wachendörfer, G., Kiefert, C. and Frost, J.W. (1982) Safety tests with Flury HEP strain 675 in wild-living European mammals. *Comp. Immunol. Microbiol. Infect. Dis.*, **5**, 177–80.

Wandeler, A., Wachendörfer, G., Förster, U., Krekel, H., Schale, W., Müller, J. and Steck, F. (1974) Rabies in wild carnivores in central Europe I. Epidemiological studies. *Zentbl. VetMed. B*, **21**, 735–56.

Waterhouse, R. (1971) *Report of the Committee of Inquiry on Rabies. Final Report.* HMSO, London.

Westergaard, J.M. (1982) Measures applied in Denmark to control the rabies epizootic in 1977–1980. *Comp. Immunol. Microbiol. Infect. Dis.*, **5**, 383–7.

Twenty-one

The mink menace? A reappraisal

N. DUNSTONE AND M. IRELAND

SUMMARY

The pest status of mink is considered in relation to their predation on domestic stock, potential impact on wildlife through predation and competition, and general nuisance value to man.

Demographic data are presented for a coast-living population of mink studied over a 6-year period. Life-history parameters including residency, fecundity and dispersal are examined. Behavioural factors affecting trappability are described and the efficacy of control procedures assessed.

Diet is discussed in relation to predation on economically important species and an attempt is made to estimate the ecological impact on the principal prey species.

INTRODUCTION

It is now some 60 years since American mink (*Mustela vison* Schreber) were first imported to Britain for breeding on a commercial scale (Thompson, 1962), and 30 years since they were first recorded breeding in the wild (Linn and Stevenson, 1980). Despite early doubts as to the viability of feral populations and a Ministry of Agriculture scheme to eradicate them (Thompson, 1962) during which period over 5000 animals were trapped, there has been a dramatic colonization of a large proportion of the waterways in the UK. The origin and spread of mink populations in the British Isles has been well documented (Thompson, 1962; Deane and O'Gorman, 1969; Cuthbert, 1973; Fairley, 1980; Lever, 1985; Birks, 1986), and it is beyond the scope of this paper to consider it further.

The pest status of the mink has been the subject of vociferous debate for nearly 10 years (e.g. Lever, 1987a,b, 1985; Linn and Chanin, 1978a,b), yet there has been little critical assessment of the extent of damage caused by this exotic carnivore, and much of what has been said refers to anecdotal accounts of depredations on domestic livestock.

In order to evaluate the pest status of the mink, we need to assess whether there has been significant economic loss or damage to domestic stock; the impact on native wildlife through predation or competition; and the nuisance value of mink to the general public. Secondly, in order to suggest a policy for the management of mink in this country we need to assess how many there are, their fecundity and dispersal capabilities and the potential for their control.

In this chapter we present an overview of current data on these various aspects of mink biology taken mainly from studies conducted in the UK, and especially information from a long-term study of mink in southern Scotland.

POPULATION DYNAMICS

Although the country has not been systematically surveyed, the presence or absence of mink has been recorded during the recent otter surveys of England and Wales and Scotland. Most watersheds in the UK now hold substantial populations of mink. The present distribution comprises the whole of Great Britain with the exception of some mountainous and urban areas (Birks, 1986). The incidence of escapes from existing fur-farms is now of minor significance in contributing to the maintenance of feral populations, with the possible exception of those instances where deliberate mass releases have occurred.

The association of mink with watercourses has been documented wherever they occur, ranging from eutrophic lakes to oligotrophic moorland rivers, coastal areas and offshore islands. Two features appear to determine habitat suitability for mink, the availability of prey and den sites (Allen, 1983). The high degree of ecological adaptability shown by mink in their native North America is thought to have pre-disposed the animal to successful acclimatization in comparable habitats and climates in Europe.

The density of mink appears to vary with the productivity of the habitat. Generally, large home ranges are associated with prey-impoverished habitats, such as oligotrophic rivers, and home ranges are considerably smaller in optimal areas, e.g. lacustrine and coastal habitats (Dunstone and Birks, 1985). The home range lengths of male mink have been shown to vary from 2.53 km on an oligotrophic river to 1.50 km of coastline. The home ranges of females are generally smaller ranging from 2.16 km on an oligotrophic river to 1.09 km on the coast (Dunstone and Birks, 1985). It is usual to find intersexual overlap of mink territories.

Apart from Gerell's (1971) study of Swedish mink there has been little work on the population biology of mink. Statistics from commercial trapping operations carried out in North America are inevitably biased towards a winter sample taken when the fur is prime, and data from fur-farms are not likely to be representative of wild populations. In the British Isles both Fairley (1980) and Chanin (1983) have published data based on post-mortem analyses of culled animals. The population parameters presented here were obtained during a live-trapping and radio-tracking study of mink conducted on a coastal study site in south-west Scotland from 1980 to 1986. Topographical details of the study area, and the techniques for capture and immobilizing mink have been presented elsewhere (Dunstone and Birks, 1985).

Trapping success varied considerably during different periods of the year (Figure 21.1). Male mink were particularly susceptible to capture during March, corresponding to the onset of the breeding season and the passage of numerous males through the study area. Trapping success was at its lowest during May and June for both sexes, and particularly so for females since none were captured during these months. This was probably due to the reduced mobility of the females while they still had dependent young. Trapping success increased during July and August as young animals became independent of their mothers and searched out territories of their own.

It was not possible to estimate the population size using mark-recapture data, since the assumptions inherent in that method were not fulfilled. However, the residential status of individuals, calculated from the amount of time elapsing between first and last capture (Figure 21.2), suggests average residency was only eight months.

Approximately 59% of the females and 40% of the males were present in the study area for longer than three months and acquired resident status. One female and eight males were classified as temporary residents, i.e. recaptured over a period of one week to three months. In addition, we

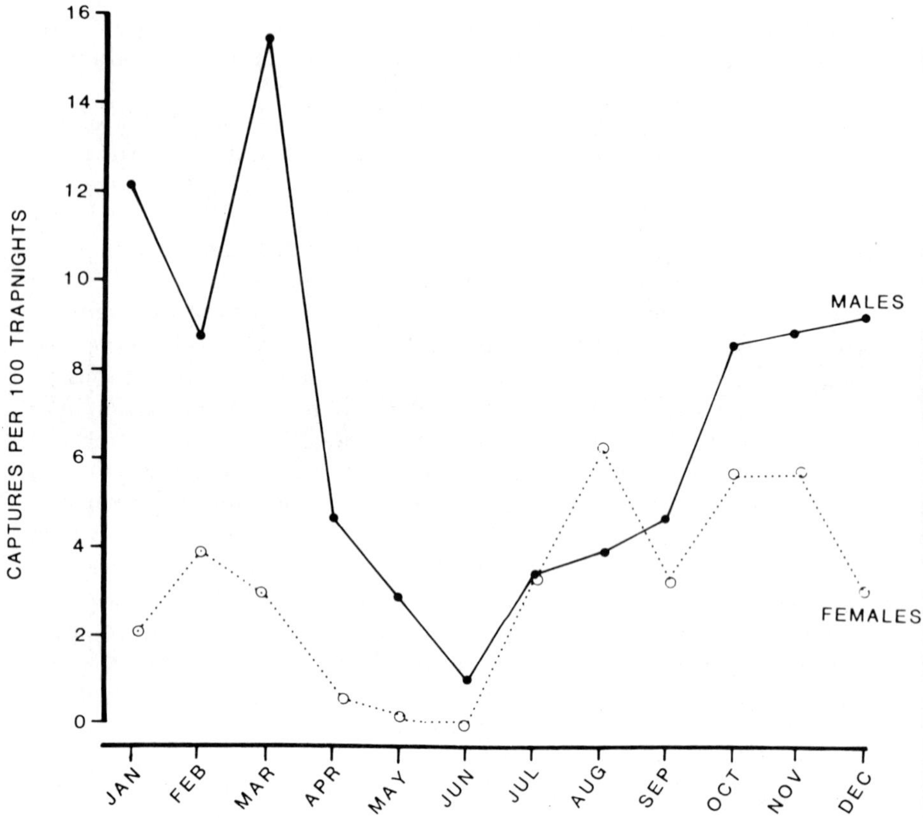

Figure 21.1 Monthly trapping success of male and female mink averaged over the five-year study.

recorded 35 transient (captured during one week only) individuals (27 males, 8 females) passing through the study area.

Table 21.1 presents data on kit production in the study area. Litters were born to two or three females each year. A mean 2.0 kits per litter survived to emergence from the den prior to dispersal. Clearly, in the wild, mink do not raise the large litters that they commonly achieve in commercial breeding operations (Dunstone, 1986).

Mink do not have many natural predators in the UK. Many mink die in their first year; the oldest animal in this study was a male of only four years. The ratio of first year animals to those two years old or older suggests an approximate mortality value of 50% per annum, although no mink were found dead from natural causes in the study area. Since on average only two kits were raised per female, juvenile mortality must be

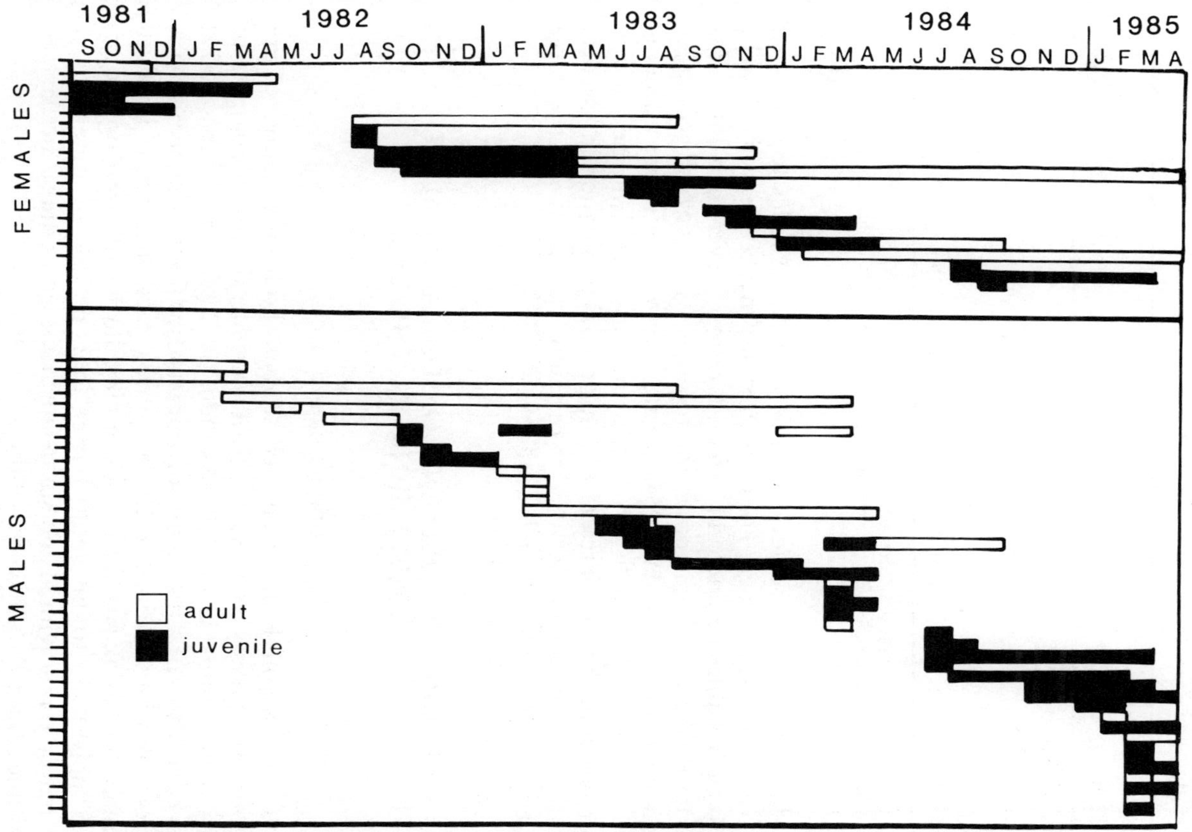

Figure 21.2 Trends in residency shown by mink over the course of study period. Juveniles assume adult status in May of the year after birth.

Table 21.1 Litter sizes of eight female mink recorded on the Ross Peninsula over a period of five years. The first figure represents estimation from counts of palpable nipples on captured females and the second value represents actual captures of kits prior to dispersal

	Female							
	6	7	10	11	18	24	25	27
1982	4/–	2/–						
1983	5/4		–/1	3/3				
1984				3/4	1/2			
1985				3/2		4/1		
1986						4/1	4/4	2/2

Mean number of kits (nipples) 3.1 per female.
Mean number of kits (captures) 2.0 per female.

considerably higher to achieve the 1 : 1 ratio of juveniles to adults we observed at one year of age. The high mortality in juveniles is likely to be due to their failure to find and defend a territory over the winter. It is evident that there is a high population turnover with a virtually complete replacement of the population occurring within three years; a finding similar to that of Mitchell (1961) for mink in North America and in agreement with that of Gerell (1971) for a feral population in Sweden. This rapid replacement is almost certainly facilitated by the large population of transient animals available.

NATURE AND EXTENT OF DAMAGE CAUSED BY WILD MINK IN THE UK

Predation on native fauna

The diet of the mink has been well studied, both in its native North America and in many countries where it is now feral. Early investigations of the diet of mink in the UK (Akande, 1972; Day and Linn, 1972) involved analysis of gut contents of animals captured during nationwide control operations during the 1960s. More recent studies have involved the analysis of mink faeces (scats), enabling the diet of local populations or of individuals to be studied over long periods of time to give information on individual and sex-related differences in the diet, and to determine seasonal changes in the diet.

Several detailed studies on the diet of feral mink have been carried out on populations inhabiting a variety of aquatic habitats in the British Isles, including riverine (Chanin and Linn, 1980), lacustrine (Chanin and Linn,

1980) and coastal habitats (Dunstone and Birks, 1987). As in studies conducted in the USA (Sealander, 1943; Korschgen, 1958) and Scandinavia (Gerell, 1967, 1968) mink in the UK have been shown to be opportunistic generalist predators exploiting a wide range of terrestrial and aquatic food items, principally mammals, fish and birds. Typically it has been found that mink respond quickly to changes in the availability of their prey, and furthermore, that the variability in the food habits of mink can depend to a large extent upon seasonal and habitat features. For these reasons it is difficult to make generalizations about mink diet.

A summary of results from dietary studies is presented in Table 21.2. Data are only presented for those prey species which have been identified as being potentially at risk from mink, e.g. ground-nesting birds (Anseriformes and Charadriiformes), or have some commercial value (Galliformes and salmonids). Considerable care must be used in the interpretation of these data, since the source of the sample (faecal material or den contents), type of analysis carried out and the method of presentation of the results differ between some of the studies. Nevertheless some interesting comparisons can be drawn from these data. In three studies (Akande, 1972; Cuthbert, 1979; Chanin and Linn, 1980) salmonids (*Salmo* spp.) were the predominant prey item, accounting for 46.6, 31.0 and 34.2% frequency of occurrence in the diet respectively. The only coarse fish species recorded in any quantity has been the eel (*Anguilla anguilla*), 11–23%; (Cuthbert, 1979; Wise, Linn and Kennedy, 1981).

In no one study were Galliforme, derived either from game bird or domestic fowl, found to be greater than 5.4% occurrence (Chanin and Linn, 1980). Ground-nesting birds such as waterfowl have been shown to be at risk during their breeding season and are preyed on in marsh and wetland habitats in the USA (Sargeant, Swanson and Doty, 1973; Eberhardt and Sargeant, 1975). In the British Isles, waterfowl (Ralliformes and Anseriformes) do not appear to be as important. Yet these prey groups have been repeatedly suggested (Lever, 1978a, b, 1985) as being major constituents of the mink's diet.

It can be seen that in five out of ten studies the most frequently taken item was lagomorph prey. Rabbits, when common in the study area, are the most heavily exploited mammal, and often the most important food overall in terms of their bulk contribution to the diet (Jenkins and Harper, 1980; Wise, Linn and Kennedy, 1981; Dunstone and Birks, 1987).

Two additional points need to be considered when assessing pest status from dietary studies. First, population studies on prey have tended to be carried out retrospectively, usually some time after the presence of mink has been determined. Secondly, most studies have averaged the dietary information across individual mink in the population, and over long

Table 21.2 Occurrence of particular prey species in the diet of mink in the British Isles

Location	Habitat	Lag.	Charad.	Gallif.	Rallif.	Anser.	Salmonid.	Sample	Analysis	Author
Scotland	General	2.3	1.1	2.3	3.4	8.0	46.6	55 guts/ 33 scats	O	Akande, 1972
England	General	12.7	<1.0	5.4	11.8	2.0	12.7*	204 stomachs	O	Day and Linn, 1972
Scotland	Riverine	<1.0	<1.0	—	16.4	4.4	31.0	722 scats	O	Cuthbert, 1979
Devon	Riverine	37.8	—	—	<5.0	<5.0	22.0	448 scats	B	Wise *et al.*, 1981
Devon	Riverine	6.3	—	<1.0	—	<1.0	34.2	475 scats	O	Chanin and Linn, 1980
Devon	Riverine	<1.0	—	<1.0	16.4	4.4	4.9	153 scats	O	Chanin and Linn, 1980
Ireland	River/lake	3.8	2.0	<1.0	8.5	5.0	2.4	2510 scats	O	Ward *et al.*, 1986
Devon	Lacustrine	9.7	—	—	15.3	5.6	1.4	57 scats	O	Chanin and Linn, 1980
Devon	Lacustrine	15.4	—	—	20.0	12.0	<1.0	513 scats	B	Wise *et al.*, 1981
Scotland	Coast/agric	34.3	5.6	<1.0	—	<1.0	<1.0	2043 scats	O	Dunstone and Birks, 1987
Scotland	Coast/agric	52.1	26.0	2.1	—	—	—	96 remains	O	Birks and Dunstone, 1984

O, % occurrence; B, % bulk.
* Net weight.

periods of time. This may have the effect of masking depredations which occur in the short term or by particular individuals.

At the individual level, some mink in some populations may specialize on particular prey types for several months at a time, and could then have

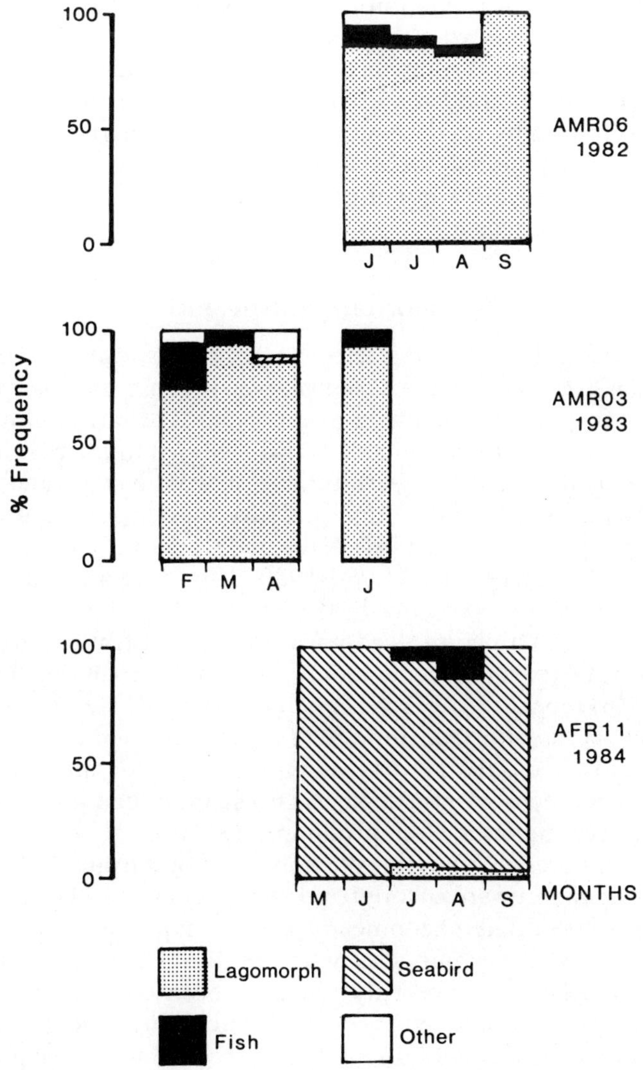

Figure 21.3 The percentage frequency occurrence of particular prey in the individual diets of two male (AMR06, AMR05) and one female (AFR11) mink on the Ross Peninsula, southern Scotland.

a considerable local impact on these species. Figure 21.3 shows selected individual dietary results from two male and one female mink inhabiting coastal habitat. The faecal material on which these results are based was collected regularly from the dens they occupied. The two male mink direct their predation towards lagomorph prey, and were known from concurrent radio-tracking studies to be resident in rabbit burrows at this time. Their pattern of predation coincides with the peak availability of rabbit prey on this study area (Dunstone and Birks, 1987). During her three breeding seasons the female raised litters on a small offshore island (*c.* 10 ha) where her food requirements and those of her kits prior to dispersal were met by predation upon a colony of herring gulls (*Larus argentatus*) which nested there. Seabird carrion was plentiful on the island, and a considerable quantity remained untouched by mink at the end of the breeding season.

Predation on domestic stock

The occurrence of prey remains derived from domestic animals in the diet of mink is an extremely rare occurrence, even in those studies which have involved mink populations living in the vicinity of farm buildings and game-rearing pens (Dunstone and Birks, 1987). Initial fears as to the importance of mink as an agricultural pest were exaggerated. Studies conducted over long periods of time and involving large samples of scats, show that poultry and game birds make up less than 1% of the diet (e.g. Cuthbert, 1979; Chanin and Linn, 1980; Dunstone and Birks, 1985). Nevertheless mink do take game birds: Fairley (1980) reports one female mink being shot in a pen of dead pheasants and of feathers from game birds occurring in some mink stomachs. Similarly Ward, Small and Fairley (1986) report the occurrence of poultry feathers in 0.2% of their scat sample. Birks and Dunstone (1985) also report the occurrence of single partridge (*Perdix perdix*) and pheasant (*Phasianus colchicus*) carcasses in the dens of mink. Nevertheless the predation of mink on Galliforme prey species is often presumptive.

Many carnivores exhibit the phenomenon of surplus killing under conditions where prey are encountered at high density and cannot escape from the predator. The phenomenon is not unique to mink, but tends to receive considerable publicity when it does occur. Surplus killing incidents are unusual in the wild but have been reported from North America (e.g. Errington, 1961). The habit of caching food has occasionally been noted in the wild and was interpreted as an adaptive trait, conferring an advantage upon opportunistic predators which exploit food resources of varying abundance and availability.

ECOLOGICAL IMPACT OF MINK ON PREY POPULATIONS

All considerations of the diet of mink should be viewed in relation to their ecological impact in terms of the numbers of prey animals they are taking and whether or not these are 'sensitive species'. The biological value of a single prey item is a function not only of its nutritional quality, but also of its abundance relative to other prey types, and the energy expended in search, pursuit, capture and handling. The relative digestibility of different prey types and the proportion of indigestible remains, e.g. bone, fur, feathers, will also vary but these may need to be taken in to provide roughage in the diet.

Information is available on the daily energy requirements of the mink and data from laboratory feeding trials allow the value of the different prey types to be assessed. From this information it is possible, albeit approximately, to estimate the mink's daily prey requirement and hence their annual impact.

The estimated weight specific energy requirements of mink range from 635 kJ/kg$^{0.75}$/day (Cowan, Wood and Kitts, 1957) to 1299 kJ/kg$^{0.75}$/day (Gregory, 1987). Birks and Dunstone (1985) report the mean weight for wild mink to be 1195 g and 688 g for males and females respectively, thus, using Gregory's (1987) values, this would suggest a daily energy requirement of 1484 kJ for a male and 981 kJ for a female. The size of prey taken by mink has occasionally been documented, e.g. adult rabbits (Birks and Dunstone, 1985), and the sizes of fish commonly taken by mink has also been determined, typically 6–12 cm for salmonids and 20–30 cm for eels (Chanin and Linn, 1980; Wise *et al.*, 1981).

Thus it is possible to estimate the weight of prey required to fulfil the daily energy requirement of a mink, and therefore the quantity of each prey type that would be required to sustain the predator. Table 21.3 shows the approximate daily consumption of various prey species by male and female mink. No attempt is made to take account of varying energy demands incurred by the animal at different times of the year, nor the effect of ambient temperature. The values should thus be considered as maxima but give an indication of the relative value of different prey types.

Undoubtedly the larger prey items, e.g. rabbits and large birds are more economic prey since they could sustain the mink for a number of days. Small mammals (e.g. voles) or passerine birds would have to be obtained in considerable numbers to meet the energy needs of the predator; mink could have a significant impact on their populations if they were specialist predators. In terms of energy content, fish prey provide a better yield than do terrestrial species. However, recent studies (Stephenson *et al.*, 1987)

Table 21.3 Energy content of prey types commonly taken by mink and the theoretical duration for which individual items would sustain the predator (see text for further explanation)

Prey type	Mean wet weight (g)	Calorific value kJ/g dry wt	Energy content (kJ)	Longevity of carcass Male (days)	Longevity of carcass Female (days)
Mammals					
Rabbit (*Oryctolagus cuniculus*)	1500[e]	20.69[b]	9311	6.3	9.5
Brown hare (*Lepus timidus*)	3540[e]	23.75[b]	25 222	17.0	25.7
Field vole (*Microtus agrestis*)	35[e]	21.0[a]	222	0.15	0.23
Birds					
Starling (*Sturnus vulgaris*)	75[f]	24.33[a]	547	0.37	0.56
Domestic fowl (*Gallus domesticus*)	1341[c]	23.64[c]	9828	6.62	10.02
Fish					
Eel (30 cm) (*Anguilla anguilla*)	290	12.93[c*]	1126	0.76	1.15
Trout (20 cm) (*Salmo trutta*)	113[f]	32.36[c,g]	1097	0.74	1.12

[a] Moors (1977)
[b] Myrcha (1968)
[c] Gregory (1987)
* Wet weight
[d] Sidwell (1974)
[e] Corbett and Southern (1977)
[f] Frost and Brown (1967)
[g] Murray and Burt (undated pamphlet)

have demonstrated that the capture of aquatic prey will require a considerably greater level of energy expenditure to provide an equivalent level of intake.

POTENTIAL INTERACTIONS BETWEEN MINK AND OTHER NATIVE CARNIVORES

Competition between mink and otters

Competition occurs where two species are utilizing a resource, e.g. food type or den, which is actually or potentially limiting. In the case of competition between the mink and otter the shared prey resource will in most instances be fish, of which otters are the considerably superior predator in terms of hunting efficiency. Mink are generalist predators and take a smaller proportion (20–60%) of their prey from aquatic sources than do otters (90–100%, Erlinge (1969)). A substantial proportion of the mink's diet consists of items of minor importance to the otter, e.g. mammals and birds. Fish may be as much as three times more important a dietary constituent to otters than they are to mink. Nevertheless, dietary overlap does occur and this may be of some importance during some seasons. Wise, Linn and Kennedy (1981) estimated dietary overlap to be as great as 40% and Erlinge (1969) has estimated 55–75% overlap. Greatest overlap would be expected to occur during the summer months when fish, in some rivers, are less available to aquatic predators. At this time it is generally noted (Jenkins and Harper, 1980; Wise, Linn and Kennedy, 1981; Dunstone and Birks, 1987) that mink are preying heavily upon rabbits, if available locally.

Wise, Linn and Kennedy (1981) further consider that the lack of size selection shown by otters and mink when preying on fish discounts any suggestion of competition between these two species. In those areas where otter diet was studied before the introduction of mink there has been no subsequent change recorded in the diet of the otter (Chanin, 1986).

It has been suggested (Lever, 1985) that mink have adversely affected the status of otters, but there is little evidence to support this contention. Chanin and Jefferies (1978) report that the decline in the otter population occurred before mink became widely established. The low otter population may have facilitated the rapid spread of mink, but it is unlikely that the occurrence of mink on a river will impair the ability of the otter population to become restablished. Concern that mink may prey on otter cubs seems to be without foundation; indeed it has been reported that mink may share a den with otters (Melquist, Whitman and Hornocker, 1981; Green, Green and Jeffries, 1984).

Mink and other mustelids

The only other carnivores with which the mink is likely to be in direct competition are the polecat (*Mustela putorius*) and the stoat (*Mustela erminea*). The staple item of diet for these predators is the rabbit (Day, 1968; Walton, 1968). Populations of rabbits are such that competition with mink is unlikely to be a limiting factor affecting the status of either native carnivore. Furthermore the polecat appears to be extending its range even in areas where mink are present (Birks, 1986). This may in part be due to the minks' association with waterways, on which the polecat appears to be less dependent.

DISCUSSION

Early worries concerning the impact of mink as a significant predator of native fauna and domestic stock, or as a competitor with native carnivores seem to be without foundation. It can be seen that the relatively small size of mink and hence their low energy requirements coupled with their low density, approximately one per kilometre, means that they have a small effect overall. General conclusions from these dietary studies are that mink are not specialist predators, they exist in balance with available prey and do not appear to have had any significant affect on any one prey species. In Britain there have not been any extinctions or drastic reductions of any species of prey that can directly and unequivocably be attributed to mink. Concern continues to be expressed (Lever, 1985) over the effect of mink on salmonid stocks, but, given the mink's rather poor adaptations for swimming and diving (Dunstone, 1978), the animal should not be considered as a significant economic pest of wild fish. It should also be pointed out that the daily intake of mink is small relative to the 1–1.5 kg of fish needed to sustain an otter (Erlinge, 1968). The studies in which salmonids were important dietary items, e.g. on oligotrophic rivers, tend to be in areas where alternative prey were uncommon.

Large mammals are by far the most economic prey for the mink, and lagomorphs, particulary rabbits, are always taken if available. Water voles (*Arvicola terrestris*) were rarely found in the diet of mink. Populations of water voles continue to flourish on some rivers despite the presence of mink, whereas on others decline in water vole numbers have been observed in the absence of a sympatric mink population.

The major problem caused by mink predation concerns their effect on native bird species. Here it must be remembered that mink are territorial and the size of the territory is influenced by two factors, the habitat quality and the presence of other mink. Because of this spacing mechanism any local abundance of food or seasonal glut will only be available to the territory holders, thus allowing only a small number of animals to share the food, and preventing mink from surrounding areas gaining access. Thus, for example, on Ross Island where there were abundant ground-nesting herring gulls available for most of their breeding season only one female mink preyed on the colony because it lay within her territory. The colony has continued to expand despite the annual predation pressure from this animal (Birks, 1986).

Birks (1986) considers that many prey populations are capable of withstanding substantial mink predation if they are in their optimal habitats. Despite the relatively high predation of mink on coots and moorhens at Slapton, south Devon (Chanin and Linn, 1980) these species are still abundant there. In optimal habitats the population of mink and their Ralliforme prey appears to be in equilibrium, and it is only in sub-optimal habitats that the population may be seriously affected.

Considerable concern has been expressed over the predation by mink on domestic stock, in particular, chickens, game birds and fish. Undoubtedly mink can pose a serious threat to agricultural holdings of high-density stock (e.g. game-bird rearing pens, chicken farms and fish-rearing ponds), as can many other predators, e.g. fox, otter, dog and cat. Mink are frequently blamed for surplus killing of domestic prey, usually when encountered in confined conditions. These incidents can be guarded against by good husbandry, which will, additionally, provide protection against native predators. The maintenance of traps in the vicinity of pens is all that is usually required to control mink in these instances.

In general terms, detailed study has revealed that mink predation has had very little overall effect upon prey populations in mainland Britain, although some local reductions of prey may have occurred in sub-optimal habitats (Chanin, 1981; Birks, 1986). It is doubtful if the agricultural or ecological impact of mink on domestic stock or native prey species is sufficient to warrant expenditure on widespread control measures. Protection of particularly vulnerable species in particular areas by means of local trapping campaigns against mink conducted at an appropriate time of the year are to be encouraged, and the development of mink farms in ecologically or agriculturally sensitive areas, e.g. offshore islands, should be limited.

ACKNOWLEDGEMENTS

This work was supported by a research grant from the Natural Environment Research Council to N.D. and a studentship to M.I. We gratefully acknowledge the assistance and humour of Dr J.D.S. Birks during fieldwork; Dr. J. Gregory for useful comment and allowing access to his unpublished results on mink nutrition; and to Dr I. Albyrak and Paul Loftus for assistance with the analysis of mink scats. We are grateful to Mr B. Finlay for permission to work on the Ross Peninsula.

REFERENCES

Akande, M. (1972) The food of feral mink (*Mustela vison*) in Scotland. *J. Zool., Lond.*, **167**, 475–9.

Allen, A.W. (1983) Habitat suitability index models: mink. *US Dept. Int. Fish. & Wld. Ser.* OBS-82/10.61.

Birks, J.D.S. (1986) *The Mink*. Anthony Nelson. Oswestry, England, 24 pp.

Birks, J.D.S. and Dunstone, N. (1984) A note on the prey remains collected from the dens of feral mink (*Mustela vison*) in a coastal habitat. *J. Zool. Lond.*, **203**, 279–81.

Birks, J.D.S. and Dunstone, N. (1985) Sex-related differences in the diet of mink *Mustela vison. Holarctic Ecol.*, 8, 245–52.

Chanin, P.R.F. (1981) The diet of the otter and its relations with the feral mink in two areas of southwest England. *Acta Theriol.*, **26**, 83–95.

Chanin, P.R.F. (1983) Observations on two populations of mink in Devon, UK. *Mammalia*, **47**, 463–76.

Chanin, P.R.F. (1986) *The Natural History of Otters*. Croom Helm. London.

Chanin, P.R.F. and Jefferies, D.J. (1978) The decline of the otter *Lutra lutra* in Britain: an analysis of hunting records and discussion of causes. *Biol. J. Linn. Soc.*, **10**, 305–28.

Chanin, P.R.F. and Linn, I.J. (1980) The diet of the feral mink (*Mustela vison*) in southwest Britain. *J. Zool., Lond.*, **192**, 205–23.

Corbett, G.B. and Southern, H.N. (1977) *The Handbook of British Mammals*. Blackwell, Oxford.

Cowan, I. Mc. T., Wood, A.J. and Kitts, W.D. (1957) Feed requirements of deer, bear and mink for growth and maintenance. *Trans. 22nd North Am. Wld. Conf.*, **22**, 179–88.

Cuthbert, J.H. (1973) The origin and distribution of feral mink (*Mustela vison*) in Scotland. *Mammal Rev.*, **3**, 97–103.

Cuthbert, J.H. (1979) Food studies of mink (*Mustela vison* Schreber) in Scotland. *Fish Mgmt.*, **10**, 17–25.

Day, M.G. (1968) Food habits of British stoats (*Mustela erminea*) and weasels (*Mustela nivalis*). *J. Zool., Lond.*, **155**, 485–97.

Day, M.G. and Linn, I.J. (1972) Notes on the food of feral mink (*Mustela vison*) in England and Wales. *J. Zool., Lond.*, **167**, 463–73.

Deane, C.D. and O'Gorman, F. (1969) The spread of feral mink in Ireland. *Irish Nat. J.*, **16**, 198–202.

Dunstone, N. (1978) The fish-catching strategy of the mink (*Mustela vison*); time-budgeting of hunting effort? *Behaviour*, **67**, 157–77.

Dunstone, N. (1986) Exploited animals: the mink. *Biologist*, **33**, 69–75.

Dunstone, N. and Birks, J.D.S. (1983) Activity budget and habitat usage by coast-living mink (*Mustela vison* Schreber). *Acta Zool. Fenn.*, **174**, 181–91.

Dunstone, N. and Birks, J.D.S. (1985) The comparative ecology of coastal, riverine and lacustrine mink *Mustela vison* in Britain. *Z. fur Ange. Zool.*, **59**, 59–70.

Dunstone, N. and Birks, J.D.S. (1987) The feeding ecology of mink (*Mustela vison*) in coastal habitat. *J. Zool., Lond.*, **212**, 69–83.

Eberhardt, L.E. and Sargeant, A.B. (1975) Mink predation on prairie marshes during waterfowl breeding season. *Proc. 1975 Predator Symp.* (*Am. Soc. Mammal.*) (eds R.L. Phillips and C. Jonkel).

Erlinge, S. (1968) Food studies on captive otters *Lutra lutra*. *Oikos*, **19**, 259–70.

Erlinge, S. (1969) Food habits of the otter (*Lutra lutra*) and mink (*Mustela vison*) in a trout water in southern Sweden. *Oikos*, **20**, 1–7.

Errington, P.L. (1961) *Muskrats and Marsh Management*. University of Nebraska Press, 183 pp.

Fairley, J. (1980) Observations on a collection of feral Irish mink *Mustela vison* Schreber. *Proc. R. Irish Acad.*, **80B**, 79–90.

Frost, W.E. and Brown, M.E. (1967) *The Trout*. Collins, London.

Gerell, R. (1967) Food selection in relation to habitat in mink *Mustela vison* in Sweden. *Oikos*, **18**, 233–46.

Gerell, R. (1968) Food habits of the mink *Mustela vison* in Sweden. *Viltrevy*, **5**, 119–211.

Gerell, R. (1971) Population studies on mink, *Mustela vison*, in southern Sweden. *Viltrevy*, **8**, 84–109.

Green, J., Green, R. and Jefferies, D.J. (1984) A radio-tracking study of otters *Lutra lutra* on a Perthshire river system. *Lutra,* **27**, 85–145.

Gregory, J. (1987) Nutritional aspects of sexual dimorphism in the American mink *Mustela vison* Schreber. PhD Thesis University of Durham.

Jenkins, D. and Harper, R.J. (1980) Ecology of otters in northern Scotland. 2. Analysis of otter (*Lutra lutra*) and mink (*Mustela vison*) faeces from Deeside, N.E. Scotland in 1977–78. *J. Anim. Ecol.*, **49**, 737–54.

Korschgen, L.J. (1958) December food habits of the mink in Missouri. *J. Mammal.*, **39**, 521–7.

Lever, C. (1978a) Are wild mink a threat? *New Scientist*, **80**, 712.

Lever, C. (1978b) The not so innocuous mink? *New Scientist*, **78**, 812–14.

Lever, C. (1985) *Naturalised Animals of the World.* Longman, London.

Linn, I.J. and Chanin, P.R.F. (1978a) Are mink really pests in Britain? *New Scientist,* **77**, 500–562.

Linn, I.J. and Chanin, P.R.F. (1978b) More on the mink 'menace'. *New Scientist,* **80**, 38–40.

Linn, I.J. and Stevenson, J.H.F. (1980) Feral mink in Devon. *Nature in Devon,* **1**, 7–27.

Melquist, W.E., Whitman, J.S. and Hornocker, M.G. (1981) Resource partitioning and coexistence of sympatric mink and otter populations in *Proc. Worldwide Furbearer Conf.* (eds J.A. Chapman and D. Pursley), Frostburg, pp. 187–220.

Mitchell, J.L. (1961) Mink movements on a Montana river. *J. Wildl. Mgmt,* **25**, 48–53.

Moor, P. (1977) Studies of the metabolism, food consumption and assimilation efficiency in the weasel (*Mustela nivalis*). *Oecologia,* **27**, 185–202.

Murray, J. and Burt, J.R. (undated pamphlet) Composition of fish. MAFF, Torrey Res. Sta. Advis. Note 38.

Myrcha, A. (1968) Calorific value and chemical composition of the body of the European hare (*Lepus europaeus*). *Acta Theriol.*, **13**, 65–70.

Sargeant, A.B., Swanson, G.A. and Doty, H.A. (1973) Selective predation by mink on waterfowl. *Am. Midl. Nat.*, **89**, 208–14.

Sealander, J.A. (1943) Winter food habits of mink (*Mustela vison*) in southern Michigan. *J. Wildl. Mgmt,* **7**, 411–17.

Sidwell, V.D. (1974) Composition of edible portion of crustaceans, fin fish and molluscs. *Mar. Fish. Rev.*, **36**, 21–35.

Stephenson, R., Butler, P.J., Dunstone, N. and Woakes, A.J. (1987) Heart rate and gas exchange in freely diving American mink (*Mustela vison*). *J. Exp. Biol.*, **134**, 435–42.

Thompson, H.V. (1962) Wild mink in Britain. *New Scientist,* **13**, 130–2.

Walton, K.C. (1968) Studies on the biology of the polecat *Putorius putorius* in Great Britain. MSc Thesis. University of Durham.

Ward, D.P., Small, C.M. and Fairley, J.S. (1986) The food of mink *Mustela vison* in the Irish midlands. *Proc. R. Irish Acad.*, **86B**, 169–82.

Wise, M.H., Linn, I.J. and Kennedy, C.R. (1981) A comparison of the feeding biology of mink (*Mustela vison*) and otter (*Lutra lutra*). *J. Zool., Lond.*, **195**, 181–213.

Twenty-two

Economic damage by feral American mink (*Mustela vison*) in England and Wales

M.D.K. HARRISON AND R.G. SYMES

SUMMARY

The importance of economic damage by the introduced predator *Mustela vison* in England and Wales was appraised using reports of damage collected by Ministry of Agriculture, Fisheries and Food advisers and a survey of trout farmers. Damage to poultry was numerically the most frequent problem but incidents were individually minor. Predation of reared game birds may be more costly but does not appear to be a significant problem. Mink were reported by 46% of trout farmers but only 16% regarded this as a serious problem; damage to fish cages and scratching/stress of trout caused concern.

Prevention of economic damage by trapping and proofing appear to be satisfactory solutions to most types of damage and it is concluded that mink are not having any seriously detrimental effects at present although numbers appear to be still increasing.

INTRODUCTION

The ecological and economic impact of the introduction of feral American mink (*Mustela vison* Schreber) in Britain has been the subject of much controversy (see Linn and Chanin, 1978; Lever, 1978; Chanin, 1981, for diverse views). Predation on a wide variety of species has been reported (Day and Linn, 1972), including domestic livestock and sporting quarry such as game birds and fish (Thompson, 1962). Mink have been present in the wild for at least 30 years (Linn, 1958; Thompson, 1962), during which time a body of experience on economic damage and its prevention has accumulated. However, no systematic analysis of types or extent of damage caused by mink has been published, and the public image of the species is coloured by anecdotal and often exaggerated accounts in popular magazines and newspapers. We review here available information on the impact of mink on economic interests (including game rearing) but do not deal with their impact on natural ecosystems (see Chapter 21). At present, their colonization appears to be still incomplete (Birks, 1986), indicating that a final balance between man's activities and the species' potential has probably yet to be reached.

The information presented here is based mainly on the reports of the Ministry of Agriculture, Fisheries and Food (MAFF) advisers for the years 1985 and 1986, reports of damage caused by mink made to MAFF for the period 1962 to 1970 inclusive, and a survey of fish farmers carried out in 1987. MAFF has been closely involved with feral mink and their control since the early 1960s (Thompson, 1968), and through its regional network of wildlife advisers (who also act in Wales on behalf of the Welsh Office) is the main source of advice on mink control in England and Wales. The Destructive Imported Animals Act 1932 requires occupiers to report to MAFF the presence of feral mink on their land, and occasionally such reports are received; however, most reports reflect occupiers' concern and a requirement for advice over potential or actual damage to livestock, game or wildlife.

METHODS

Individual notifications to MAFF of damage attributed to feral mink were extracted from MAFF files for the period 1962 to 1970 for South Western England (the area contained at that time in the counties of Gloucestershire, Wiltshire, Dorset, Somerset, Devon and Cornwall). After 1970, the reduced input by MAFF into mink control led to fewer reports of damage being made.

For 1985 and 1986 requests for advice to MAFF Wildlife and Storage Biologists (WSB) were recorded on WSB's national database. Details were available of date, map reference, commodity damaged, and type of enquirer (e.g. farmer, gamekeeper, etc). The species recorded was that believed by the adviser to be responsible for the damage.

A telephone survey of trout farmers in England and Wales was carried out in September 1987. Farms contained in a membership list supplied by the British Trout Association were asked questions to elucidate their awareness of the presence of mink, their evaluation of its significance, and control efforts taken. A total of 75 farmers were contacted, all of whom agreed to answer questions.

RESULTS AND DISCUSSION

Numerically, predation on domestic poultry and water fowl was the most frequent problem notified, and this has remained roughly the same as a proportion of total reports (58% in the 1960s, against 61% in the 1980s). Reported damage to wild birds and fish (freshwater fisheries) were next in order of frequency. It has been argued (Linn and Chanin, 1978) that in some circumstances the impact of mink is neutral or beneficial, although in others, e.g. 'put-and-take' fisheries, damage may be real. Damage to fisheries is not discussed further because it is outside the scope of this

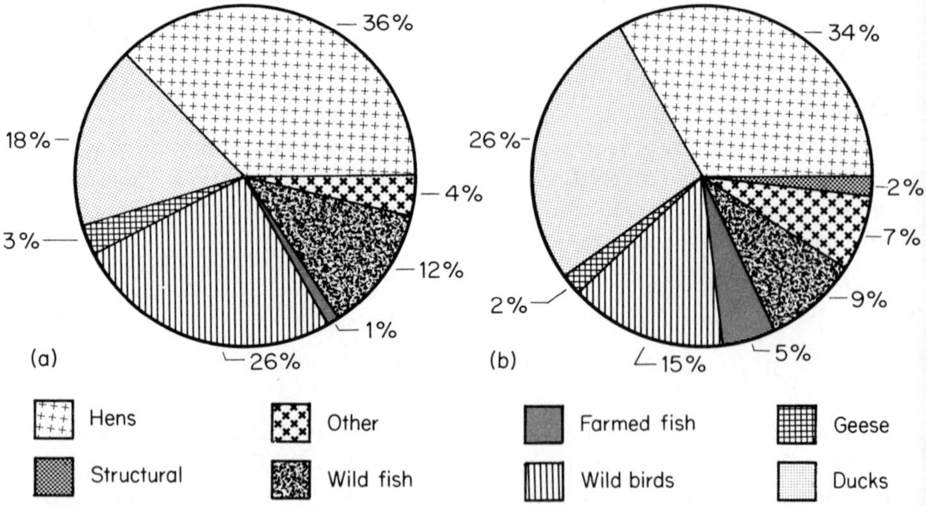

Figure 22.1 Frequencies of different types of damage caused by mink notified to MAFF in south west England in (a) 1962–1970 (n = 108) and (b) 1985–1986 (n = 87).

chapter, and is unquantifiable. Damage to farmed fish represented 1% of reports in the 1960s, against 5% in 1985/1986; this increase is not statistically significant.

Figures 22.1(a) and (b) suggest there has been little change in the types and proportion of damage experienced in south-west England in the period following the introduction of mink. Changes in rural industries (for example, the rapid increase in the number of trout farms) over this period may have had more effect on the types of damage experienced than any changes in the behaviour of mink following their establishment in the wild.

Mink predation on poultry and domestic water fowl

The killing of domestic poultry, ducks and geese has been widely recorded (Chanin, 1976, 1981; Clark, 1970; Lever, 1977; Thompson, 1962) although it has often been assumed to be insignificant in economic terms (e.g. Linn, 1984).

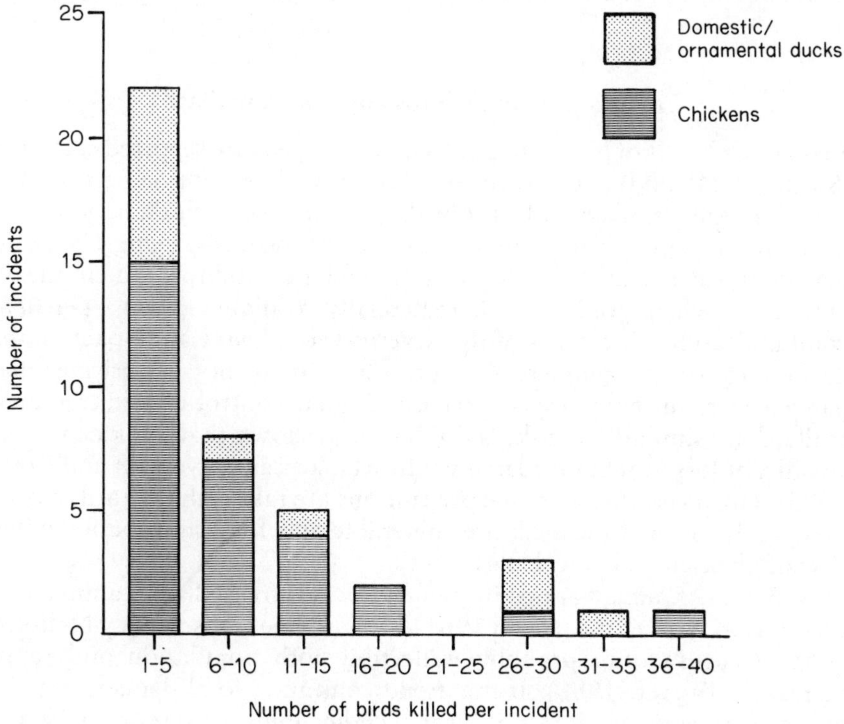

Figure 22.2 The number of birds killed (where recorded) in individual cases of mink predation on poultry and water fowl (1962–1970 and 1985–1986).

Figure 22.2 shows numbers of birds killed to be generally low, despite a probable bias for the more serious cases of predation to be reported to MAFF. A typical domestic laying hen is worth approximately £2 (although some show birds may be significantly more valuable), so it is clear that most incidents of mink predation on poultry do not individually represent a serious financial loss. Reports of mink damage to poultry were most frequent in the spring and especially late summer (August/ September), the latter perhaps due to predation by dispersing juvenile mink.

Although individual incidents of predation are relatively insignificant, mink is the species most frequently (53% of reports) associated with requests for advice on predator damage to poultry. Foxes comprised 35% of reports, badgers 7%, stoats, weasels, polecats and ferrets 1% each: $n = 416$ (England and Wales 1985–1986; WSB computerized reports). These figures are probably biased since poultry-keepers are likely to be more familiar with solutions to fox damage, and hence less likely to seek advice on its prevention; but they suggest that mink rank highly on the list of mammalian predators of poultry.

Predation of game birds and reared mallard

Several mass kills of pheasant poults in release pens are reported each year (Swan, 1984) and a few such instances have been notified to MAFF, including kills in excess of 100 birds. The cost of such damage would normally be equal to the replacement cost of the birds, being £2–£2.50 per poult, since nearly all kills occur immediately following their release at a time when poults are commercially available (Swan, personal communication). Incidents of this severity would have an impact on the economics of a commercial shoot, and it is not surprising that gamekeepers in many areas carry out regular control of mink. Reared mallard are similarly at risk, and it has been shown that mink can cause significant losses of hand-reared wildfowl (Sargeant, Swanson and Doty, 1973). However, if reasonable precautions are taken, the hazard may be acceptably low and the birds are vulnerable only for a short period in the case of pheasants (Swan, 1984).

Numbers of mink trapped by gamekeepers increased at an annual rate of about 21% over the period 1961–1985 (Game Conservancy National Game Census data), probably reflecting both population and range increases (Tapper, 1980). If this trend continues, total damage may be expected to rise; but it is not clear whether mink predation on reared game is likely to become a common problem.

Predation at trout farms

Mink predation at fish farms has been reported by Dennis (1979) in Scandinavia, where the introduction of feral American mink parallels but probably pre-dates the situation in the British Isles (Wildhagen, 1956). Trout farms provide a source of abundant food for feral mink, and they are often sited in mink habitat due to their need for high quality water supplies.

Of the 75 surveyed trout farmers 46% said they had been aware of mink visiting their farms in the previous year. Of these, 34% (16% of the total sample) regarded mink damage as a serious problem. Types of damage described by trout farmers (numbers in brackets) included: predation of fish (24), damage to fish (13), damage to fish cages/netting (3), general (2). Several farmers reported that damage (scratches/bites) was liable to initiate fungal diseases, and one farmer reported mass deaths of fish, likely to be due to stress or panic caused by a mink's attempts to make a capture. A range of trout sizes was taken, from fingerlings up to fish of 0.5 kg or more, and larger fish may be damaged by unsuccessful attempts at predation.

A majority (75%) of the farmers who regarded mink damage as serious also mentioned types of damage other than the taking of fish – stress, scratching of fish, or damage to cages – indicating that direct losses are not their only (or possibly the major) concern.

Some form of mink control was carried out by 48% of all surveyed trout farmers whereas 86% of farmers who had been aware of mink on their farms in the previous year were carrying out control. Methods of control (numbers in brackets) included: trapping (cage or spring traps) (17), cage trapping (13), shooting (7), and control by the local mink hunt (3). No farmers had carried out proofing specifically against mink, although several reported that they were not vulnerable to mink damage because fish were kept under cover or in free-standing tanks.

Figure 22.3 shows the farmers' perceptions of the relative importance of mink as predators. Total responses exceed the number of farmers in the survey because some farmers gave more than one species; herons (*Ardea cinerea*) represented 63% of responses, against 7% of responses listing mink.

Mink were reported from trout farms in most parts of the country. Most farms which trapped mink caught under six per year, although one farm trapped about 30 per year, and another had shot 40 in the previous 18 months.

It seems probable that most farmers were not over-concerned about mink damage because an effective and relatively cheap control method

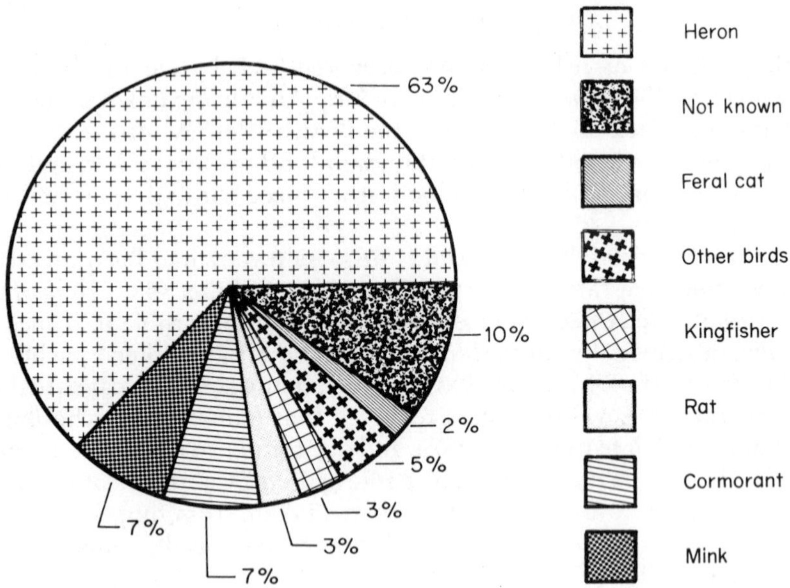

Figure 22.3 The species named by all trout farmers in the survey as their most serious predator problem ($n = 92$).

(cage trapping) is available, and they would expect to trap a mink before it had caused much damage. However, in a badly-managed unit, losses may be significant; a Cornish fish farmer claimed to have lost 500 50–75 g trout in the space of a month, after which he trapped 25 mink within a week, and a total of 40 in the period of a month (Bell, personal communication). This incident occurred in August when juveniles are dispersing (Gerell, 1970).

CONCLUSIONS

It seems likely that, as mink colonize remaining areas of suitable habitat, the British population and probably the national cost of mink damage will increase. Therefore, the position described here should be taken as an interim one; as mentioned earlier, there is evidence that mink numbers as revealed by gamekeepers' captures have increased rapidly over the last 25 years, with no sign as yet of stabilization.

A variety of aspects of mink damage and the frequency with which they appear to occur have been described. Mink seem to cause relatively low financial losses to poultry keepers, and those affected are mainly small, low-output units run as a sideline; since the 1950s most commercial poultry-keeping has taken place indoors in well-maintained buildings which should be mink-proof. The recent increase in popularity of free-range poultry-keeping has not led to reports of mink problems. However, any system of production where the housing cannot easily be made mink-proof (for example, some turkey units) will be vulnerable to predation.

Of the other main areas of damage described here, incidents of predation on game birds seem to be infrequent but may be damaging to the individual holding; by contrast, regular damage is reported on trout farms. So, to summarize these conclusions in general terms, what can we say about the status of feral mink as an economic problem?

To put the damage described in context, it is necessary to consider first what are the different economic costs imposed by a predator. Direct losses are the most obvious, but control costs may also be significant if a cheap and effective method of control is not available. Finally, a species may impose a hidden cost through having restricted the circumstances under which an enterprise may operate profitably, thus acting as a constraint on the size of that industry. These factors are interrelated; rational pest control predicts that farmers should carry out control until the unit cost of control exceeds that benefit in damage reduction which it would achieve. Where a control method is reasonably effective, expenditure on control should therefore give an indication of farmers' views of likely levels of damage; so control costs may thus provide a clearer picture of perceived levels of damage in cases (such as when species enjoy legal protection) where objective information is not available.

In the case of mink no major costs, whether through damage, control costs or hidden constraints, have been identified, so judged by these criteria it seems reasonable to conclude that feral mink are as yet a relatively minor problem; although they may be of more significance in isolated cases. As a result, it is difficult to justify on economic grounds more than local attempts at control.

A final aspect of this survey is the social impact of feral mink on farmers and others who are concerned with pest control. In several respects, the advent of the mink has meant a return to traditional management techniques which would have been carried on before the elimination of the polecat and the decline of the otter in much of England and Wales. Much as the mink has sometimes been argued to substitute in part for the ecological role of these species, so it seems that the human response has been to transfer to mink the control methods which would have been applied to the otter and polecat when these were held to be pest species. Two examples of this are the switch from otter hunting to mink hunting, and the need to proof poultry housing in the presence of mink.

ACKNOWLEGEMENTS

We thank especially the many MAFF staff and fish farmers who supplied reports of mink damage, and Michael Adams and Claire Stride for help with data collection and analysis; also the Game Conservancy for permission to use data from the National Game Census. Morris Gosling, Mike Swan, Dick Lincoln, Mark Temple and Peter Day provided helpful comments on an earlier version of this paper.

REFERENCES

Birks, J. (1986) *Mink*. Anthony Nelson.

Chanin, P. (1976) *The Ecology of Feral Mink in Devon*. PhD Thesis. University of Exeter.

Chanin, P. (1981) The feral mink – natural history, movements and control. *Nature in Devon*, **2**, 33–54.

Clark, S.P. (1970) Field experience of feral mink in Yorkshire and Lancashire. *Mammal Rev.*, **1**, 1–8.

Day, M.G. and Linn, I. (1972) Notes on the food of feral mink (*Mustela vison*) in England and Wales. *J. Zool. London*, **167**, 463–73.

Dennis, R.H. (1979) RSPB Study Tour Report: Scandinavia.

Gerell, R. (1970) Home ranges and movements of the mink *Mustela vison* in southern Sweden. *Oikos*, **21**, 160–73.

Lever, C. (1977) *The Naturalised Animals of the British Isles*. Hutchinson.

Lever, C. (1978) The not so innocuous mink? *New Scientist*, **78**, 812–14.

Linn, I. (1958) Carnivores. *Bull. Mamm. Soc. British Isles*, **9**, 4–7.

Linn, I. (1984) In two minds about mink. *BBC Wildlife*, September 1984, 435–7.

Linn, I. and Chanin, P. (1978) Are mink really pests in Britain? *New Scientist*, **77**, 560–2.

Sargeant, A.B., Swanson, G.A. and Doty, H.A. (1973) Selective predation by mink on waterfowl. *Am. Midl. Nat.*, **89**, 208–14.

Swan, M. (1984) Mink kills in release pens. *Game Conservancy Newsletter*, October 1984, p. 5.

Tapper, S. (1980) The status of some predatory mammals. *Game Conservancy Ann. Rev.*, **11**, 48–54.

Thompson, H.V. (1962) Wild mink in Britain. *New Scientist*, **13**, 130–2.

Thompson, H.V. (1968) British wild mink. *Ann. Appl. Biol.*, **61**, 345–9.

Wildhagen, A. (1956) Present distribution of North American mink in Norway. *J. Mammal.*, **37**, 116–18.

Twenty-three

British seals – vermin or scapegoats?

S.S. ANDERSON, J.H. PRIME,
J. HARWOOD AND N. BONNER

SUMMARY

The assessment of whether or not a mammal species has an impact on some aspect of its environment which is also important to man, and would thus lead it to being classified as a pest, requires a critical analysis of a variety of parameters. The case of the interaction between seals and fisheries in Britain has been examined in detail, and problems elsewhere with the same species are briefly reviewed. A summary is presented of the status of seal populations in Britain, the nature of the problem, the research which has been undertaken to quantify aspects of seal damage and of the information which has been collected on diet and quantities of fish consumed. An evaluation is made of the significance of the interactions and the implications of seal management policies.

INTRODUCTION

Seals and man have had a long history of interactions, chiefly consisting, until recently, of man's predation upon seals. Now, few seals are killed as a resource for man and numbers of seals have increased markedly in certain areas during this century. Seals impinge on man's interests in a variety of ways, but principally through their impact on fisheries. Both grey (*Halichoerus grypus*) and common (*Phoca vitulina*) seals can undoubtedly be pests at fishing operations, at least on a local basis, around the UK coast. However, grey seals have been identified as more serious offenders in terms of general interactions with fisheries. Various attempts have been made in the last few decades to reduce the grey seal population of Britain. The most recent of these occurred in Scotland in 1977 and 1978, but was called off following public protests and assertions that the scientific evidence to justify such a cull was inadequate. Conservation bodies have stated that seals should not be made scapegoats for failures in fisheries management, which implies that they have been unreasonably accredited with guilt.

Here, we present a brief review of the nature of damage caused to fisheries by seals, the techniques used to investigate the problem, the results so far available, and the arguments for and against taking action against seals. This chapter is intended as a summary of the available information; more detailed accounts of the information on which this review is based can be found elsewhere. Interactions with seals are described in more detail in Harwood (1984a, b) and Harwood and Croxall (1988). The techniques used in assessing seal diet are described in Prime and Hammond (1987). Estimates of the numbers of seals in Britain are given annually in *NERC News* (e.g. NERC, 1987). Estimates of age specific mortality on commercial white fish stocks in the North Sea caused by grey seals are given in Hammond, Prime and Prime (in press). A detailed discussion of the problem in Canadian waters is covered in The Canadian Commission on Seals and Sealing (Royal Commission, 1986).

SEAL POPULATIONS IN GREAT BRITAIN

The major grey seal breeding sites in Britain are surveyed annually by the Sea Mammal Research Unit of the Natural Environment Research Council. Smaller grey seal colonies and common seal populations are surveyed when time and resources permit. Aerial photography is the main survey method used, although counts are made on the ground at sites which are regularly visited, such as the Farne Islands, or where photography is inappropriate, for instance where seals breed in caves. Most of the current information on common seal numbers is based on surveys by boat during the pupping season, but recent research on common seal behaviour indicates that the best time to survey is during the moult in late July and early August (Thompson, in press).

The stronghold of both species is in Scottish waters (Figure 23.1),

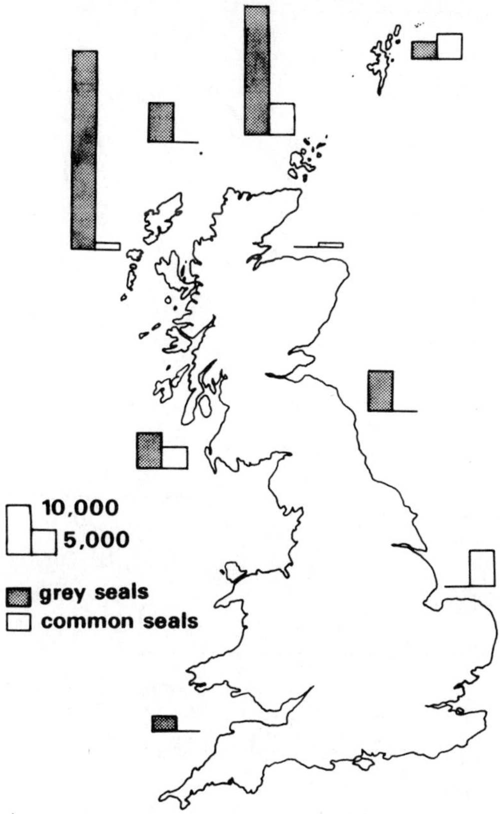

Figure 23.1 The distribution of seal populations in 1985.

although grey seals do occur on the southwest and east coasts of England, and the Wash has the largest single assembly of common seals in Britain, and probably in the world. Numbers of grey seals have been increasing since population assessment began in the early 1960s (Figure 23.2), but common seal numbers appear to have been relatively static. A probable increase has been observed in Shetland where conservation measures have reversed a decline caused by over-hunting (Anderson, 1981; NERC, 1987), and the Wash population similarly may have increased. In general, the survey methods for common seals are insufficiently precise to allow detection of anything other than gross population changes. The total size of the British grey seal population in 1985 was estimated to be 92 000 animals, while the minimum estimate of the common seal population was 25 000 (NERC, 1987).

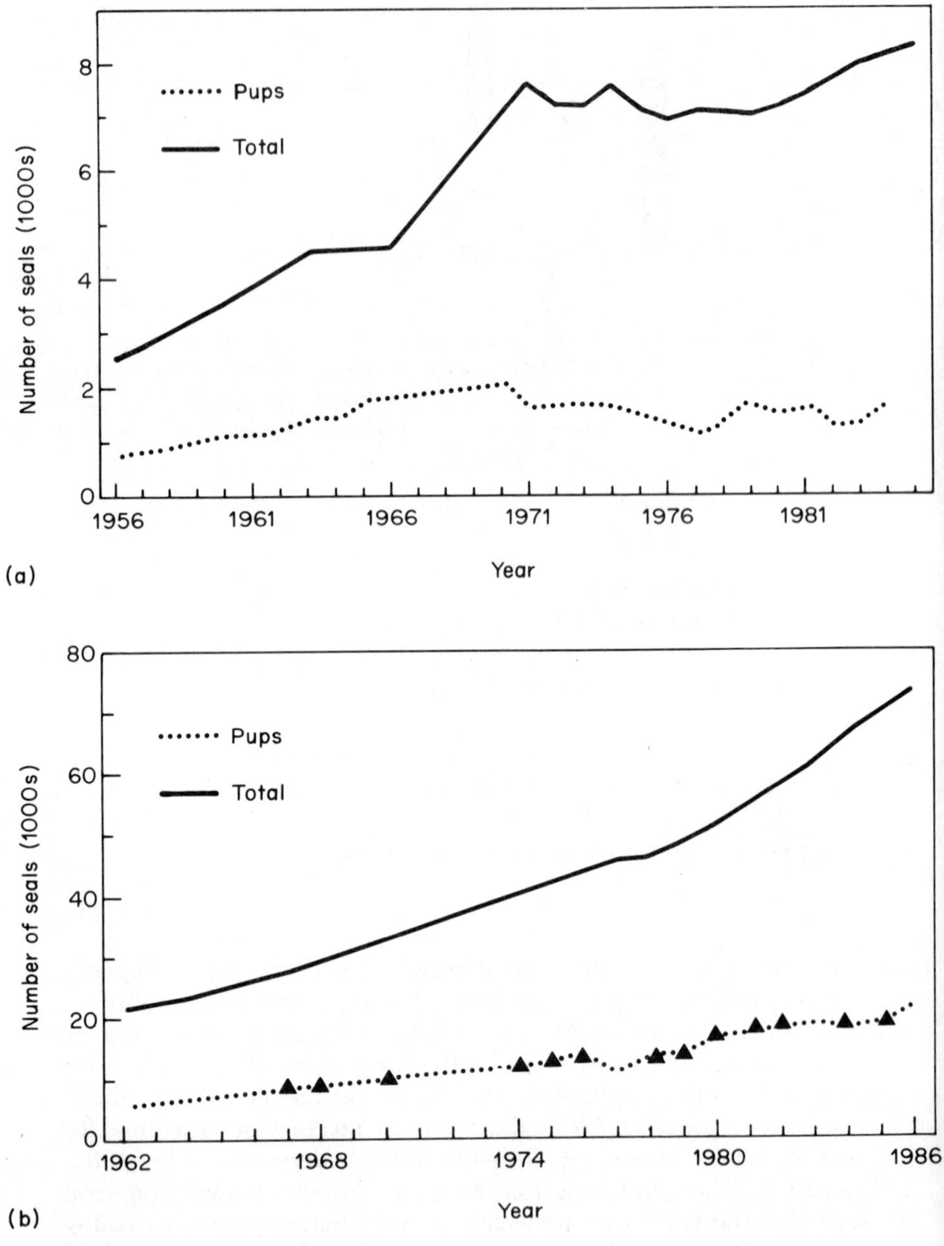

Figure 23.2 (a) Observed pup productions and estimates of total population size for the grey seal population breeding at the Farne Islands and the Isle of May. (b) Observed (▲) and estimated pup productions and estimates of total population size for the grey seal population breeding in Orkney and the Outer Hebrides.

INTERACTIONS WITH FISHERIES – THE PROBLEMS

Grey and common seals interact in three different ways with commercial fisheries.

1. They cause direct damage at netting stations and fish rearing cages.
2. They are hosts to nematode parasites which infest fish.
3. They compete with man for stocks of commercial fish.

At netting stations seals may remove fish completely from nets, damage fish caught in the nets, damage the nets themselves, become entangled and drown themselves, or drive fish away from the nets. Most of these problems occur around static gear, particularly nets set for salmon (*Salmo salar*) and tangle nets. The rearing of salmonid fish in cages moored offshore has become a growth industry around the Scottish coast in the last decade. Seals are regarded as the most serious fish predators at salmon farms, with over 80% of farms reporting some damage from seals (Ross, 1988).

Grey seals are the final host for three species of nematode parasite, *Pseudoterranova decipiens* (codworm), *Anisakis simplex* (herring worm) and *Contracaecum osculatum*, all of which use a number of commercial fish species as an intermediate host (Young, 1972). The most important of these in terms of worm burden and commercial damage is codworm. *Contracaecum* is probably the most abundant parasite in both species of seal but, as it affects the viscera of fish rather than flesh, it does not present a commercial problem. Common seals are also host to the three nematodes, but burdens are generally much lighter, and herring worm tends to dominate. This parasite has not recently presented a threat but, with the recovery of populations of herring in the North Sea, infestation levels may perhaps become a problem. The occurrence of worms in fish flesh seriously reduces its market value and increases the handling costs if removal is required. None of the parasites lead to human infection as long as the fish is properly cooked before being eaten.

The fact that fishermen see seals as competitors for the same fish stocks is probably the most important aspect of seals and fisheries interactions. This was perceived as the major reason justifying the attempted grey seal cull which began in 1977 (Parrish and Shearer, 1977). The main opposition case to the 1977/78 cull was based on the lack of scientific evidence on seal diet and consumption and the impact on commercial fisheries. Since 1978, a major effort has been invested in studying the interactions of grey seals and fisheries.

ASSESSMENT OF THE IMPACT OF SEALS ON FISHERIES

Direct damage

The interaction between seals and salmon fisheries has been studied by examining all the possible indicators of levels of interaction – salmon survival at sea, seal damage to salmon at sea, seal damage to salmon at selected netting stations along the north and east coast of Scotland and the number of seals seen and killed at these stations. A time series of data, spanning more than two decades exists for these sites. None of the indicators showed clear evidence of an increase in interaction from 1960 to 1980, a period when grey seal populations doubled (Harwood, 1984b). The proportion of salmon damaged by seals is relatively low, on average less than 5% of the catch. However, at certain times of the year, particularly early in the salmon fishing season when spring runs are low and fish relatively scarce, some netting stations do suffer much higher levels of damage. One aspect not established by the study was the origin of the seals which were causing the damage at nets. It was not clear if all members of the local seal population were equally responsible for this damage, whether it was seals moving through the area or whether it was caused by a few 'rogue' seals. The need for effective methods of control of seals at nets was stressed, but it was thought unlikely that an overall reduction in the seal population would result in a significant decrease in damage by seals at nets.

Damage by seals at salmon farms is a relatively new problem which has not been fully assessed quantitatively. One potentially effective solution to the problem lies in the construction of cages or nets which exclude seals (Ross, 1988), but at some sites anti-predator nets are difficult to set because of fast tides. Seal scaring devices appear to work initially and then habituation occurs, which is perhaps not unexpected in the presence of such a strong food stimulus (Anderson and Hawkins, 1978).

Parasites

The codworm problem is a complicated one. Levels of codworm in fish from UK waters increased considerably during the 1950s and early 1960s at the same time as the grey seal population was increasing (Rae, 1972). However, although the seal population continued to increase throughout the 1970s, codworm levels did not (Parrish and Shearer, 1977). Levels of infestation in Norway and Iceland are thought to be higher than in the UK, and in Canada there has apparently been a recent dramatic increase in codworm infection, to levels far higher than those recorded in the UK, at the same time as the grey seal population there has been rising (Royal

Commission, 1986). Thus there appear to be differences between the nature of this problem in different areas. The life history of the codworm requires at least one intermediate invertebrate host before larvae are ingested by fish and thence encyst in their musculature. Seals are host to the adult worms which take up residence in their stomachs, often causing gastrointestinal lesions (Young and Lowe, 1969). Reducing seal numbers is not likely to result in a direct reduction in levels of infestation in fish where such a complex life cycle exists. In the UK the parasite problem is not seen as a major cause for concern, but in Canada there are calls for a reduction in seal numbers for this reason (Royal Commission, 1986).

Competition for fish stocks

Although a number of different techniques have been used to estimate how much fish is eaten by different seal populations, the one used in this context is based on the daily energy requirements of the seals. With this method it is possible to compensate for the wide variations in the calorific content of different fish species and the variations between seasons of the same species. The approach is described in detail in Harwood (1984b). First the number of seals in each age-class is estimated using the results of aerial surveys of breeding colonies (see previous section on populations) and data from management culls. Then this age structure is converted into a size structure using male and female growth curves. From the size structure an estimate of the population's annual energy budget is made using measurements of the energy required for maintenance, growth and reproduction.

The contribution of different fish species to the seals' diet was calculated from an analysis of the hard parts (mostly otoliths) found in seal faeces collected at a number of haul-out sites around the British coast throughout the year. The size of these otoliths, when corrected for digestion in the seals' gut, can be used to estimate the mean size of those prey species whose hard parts are consumed by the seals and to calculate how many individuals of each species are eaten. One of the drawbacks of the faecal analysis technique is that it must underestimate the consumption of prey which leave no hard parts or whose hard parts do not come through to the faeces. The importance of prey species which are not represented by remains can be assessed by comparing the calorific content of a given volume of faecal material with an estimate of the calorific intake represented by the fish remains in the faeces. If the diet is adequately represented by the otoliths recovered, the ratio of these two energy values will lie within a range of excreted to ingested energy ratios consistent with measured values of digestive efficiency. A large discrepancy between the two ratios would indicate that the faecal samples

Table 23.1 Estimates of the quantities of fish consumed by grey seals in the North Sea compared with commercial catches

Fisheries area:	ICES Divisions IVa			ICES Division IVb, c		
Seal stock:	Orkney			Donna Nook/Farnes/Isle of May		
Species	Seal consumption (tonnes)	Commercial catch (tonnes)	%	Seal consumption (tonnes)	Commercial catch (tonnes)	%
Cod	1 250	60 917	2.1	9575	194 883	4.9
Whiting	146	35 665	0.4	1245	64 339	1.9
Haddock	690	109 341	0.6	788	65 109	1.2
Saithe	2 278	141 316	1.7	933	8 994	10.4
Norway pout				31	289 690	<0.1
Ling				374	1 182	31.6
Plaice				127	109 591	0.1
Lemon sole				407	6 344	6.4
Mackerel				21	7 583	0.3
Sandeel	29 803	73 601	40.5	5180	553 751	0.9
Dover sole				304	21 466	1.4
Flounder	535	1	>100.0	481	2 688	17.9
Dab	196	506	38.7	95	9 059	1.0

Commercial catches by all countries in the North Sea in 1982 are taken from ICES (1985).
From Harwood and Croxall (1988).

were not representative of the diet. This was not found to be the case in the grey seal studies. Another technique, currently being developed, which would indicate that prey had been taken which was not represented is that of electro-immunoassay (Grisley and Boyle, 1985). This technique examines the proteins present in the faecal material and attempts to match them with known fish species. Once the diet is determined, the mean calorific content of each species can then be used in conjunction with the population's energy requirements to calculate the actual quantities of each species consumed by seals in different areas.

The results of such calculations for seals at grey seal sites which border the North Sea are shown in Table 23.1. The data are split according to the relevant ICES divisions, so that an evaluation can be made of consumption by seals versus commercial catch by all countries fishing in the North Sea in 1982 (ICES, 1985). The split also shows the extent to which seal diet can vary from one area to another. Overall, the most important prey species for grey seals were sandeels (Ammodytidae) which comprised 61% of the diet by weight, and cod (*Gadus morhua*) which comprised 19%. Other species of fish were in all cases less than 6% of the grey seal diet. Total consumption by grey seals was estimated at 57 000 tonnes.

The ratio of the quantity of fish consumed by seals to the commercial catch is equal to the ratio of the mortality of that fish stock to the mortality caused by fishing. When compared with the fisheries data, the quantity of fish consumed by seals was generally only a few per cent of the commercial landings. In those cases where the mortality caused by seals was over 10%, such as on sandeels, flounder (*Platychthys flesus*) and dab (*Limanda limanda*) in the northern part of the North Sea and on saithe (*Pollachius virens*), ling (*Molva molva*) and flounder in the southern part, the commercial catch of the species was usually small. Sandeels are heavily exploited for fish meal, but have a very low landed value of between 1 and 10% of most commercial fish species.

WHAT EFFECT WOULD REDUCTION OR REMOVAL OF SEALS HAVE ON COMMERCIAL CATCHES?

A simplistic assumption would be that all or part of the fish not eaten by seals would be available to the fishermen. Cost/benefit analyses made on this basis (Royal Commission, 1986) suggest that a cull of grey seals in Canadian waters would be financially justified, but recent criticism of this conclusion suggests that the possible benefits of culling were overestimated tenfold (Holt and Lavigne 1988). The effect of changes in the seal population on fisheries catches will depend on the response of other predators, as well as the fish stocks themselves, to changes in levels of predation. There is no simple and reliable way to predict these. Similar caveats were made above about the difficulty of predicting the effects of seal control on infestation levels of codworm in fish where a complex life cycle exists.

A decision to cull seals is inevitably a political one, but, with recent improvements in the techniques for assessing damage to fisheries, future decisions should be based on a knowledge of the facts and an understanding of the likely outcome of management policies.

REFERENCES

Anderson, S.S. (1981) Seals in Shetland waters. *Proc. R. Soc. Edin.*, **80B**, 181–8.

Anderson, S.S. and Hawkins, A.D. (1978) Scaring seals by sound. *Mamm. Rev.*, **8**, 19–24.

Grisley, M.S. and Boyle, P.R. (1985) A new application of seriological techniques to gut content analysis. *J. Exp. Mar. Biol. Ecol.*, **90**, 1–9.

Hammond, P.S., Prime, J.H. and Prime, P.A. (in press). Age-specific mortality inflicted by grey seals on North Sea whitefish stocks. *J. Counseil.*

Harwood, J. (1984a) Competition between seals and fisheries. *Sci. Prog., Oxf.*, **71**, 429–37.

Harwood, J. (ed.) (1984b) *Interactions between Grey Seals and UK Fisheries*. Report to the Department of Agriculture and Fisheries for Scotland, 241pp.

Harwood, J. and Croxall, J. (1988) The assessment of competition between seals and commercial fisheries in the North Sea and Antarctic. *Mar. Mammal Sci.*, **4**, (1), 13–33.

Holt, S. and Lavigne, D.M. (1988) Canada overcome by new urge to kill seals. *BBC Wildlife*, **6**, January.

ICES (1985) *Bulletin Statistique des pêches maritimes pour l'année 1982*. **67**, 17pp.

Natural Environment Research Council (NERC) (1987) Seal stocks in Great Britain: Surveys conducted in 1985. *NERC News*, March.

Parrish, B.B. and Shearer, W.M., (1977) *Effects of Seals on Fisheries*. ICES CM 1977/M : 14.

Prime, J.H. and Hammond, P.S. (1987) Quantitative assessment of grey seal diet from faecal analysis in *Approaches to Marine Mammal Energetics* (eds A.C. Huntley, D.P. Coata, G.A.J. Worthy and M.A. Castellini) Society for Marine Mammalogy, Special Publication no. 1.

Rae, B.B. (1972) A review of the cod-worm problem in the North Sea and in Western Scottish waters 1958–1970. *Mar. Res.* **2**, 24 pp.

Ross, A. (1988) *Controlling Nature's Predators on Fish Farms*. Report for the Marine Conservation Society.

Royal Commission (1986) *Seals and Sealing in Canada*. Report of the Royal Commission. Ministry of Supply and Services, Canada.

Thompson, P.M. (1989) Seasonal changes in the distribution and composition of common seal (*Phoca vitulina*) haul-out groups. *J. Zool. Lond.*, **217**, 281–94.

Young, P.C. (1972) The relationship between the presence of larval anisakine in cod and marine mammals in British home waters. *J. Appl. Ecol.*, **9**, 459–85.

Young, P.C. and Lowe, D. (1969) Larval nematodes from fish of the subfamily Anisakinae and gastro-intestinal lesions in mammals. *J. Comp. Pathol.*, **79**, 301–13.

Twenty-four

Feral cats: management of urban populations and pest problems by neutering

P.F. NEVILLE

SUMMARY

In urban areas groups of stray or abandoned domestic cats, *Felis catus*, may concentrate around resources of food and shelter. Many are adopted and regularly fed by animal lovers but, unchecked, the cats usually quickly outbreed resources. Most kittens die from starvation and disease causing much distress to animal welfarists unable to accept the sight of natural population control. Pest status derives from behaviours such as pungent territorial spraying by toms, nocturnal noise from cats fighting and calling for mates and fear of zoonoses and fleas. Destruction of a cat colony only creates a vacuum soon filled by the surrounding high density of stray or urban feral cats. Trapping, neutering and returning a number of healthy adult cats to their site has proved successful at stabilizing numbers and alleviating the pest behaviour problems provided food resources are regulated.

A long-term study of a south London cat colony managed by neutering is reported with an assessment of the success of a recent programme carried out in tourist resorts in the British Virgin Islands.

In urban areas groups of stray or abandoned domestic cats (*Felis catus*) may coalesce around resources of food and shelter and, although usually thought of as solo predators, become tolerant of the presence of other individuals. Feral cats share many corners of our urban environment, inhabiting parklands, industrial and hospital sites, docklands and areas of dereliction all over the world. They seem especially well established in tourist resorts of Mediterranean countries where availability of high quality food, either scavenged or begged from tourists, and the mild climate, encourage high fecundity and good survival of young. The most recent survey estimated the population of feral cats in the UK to be about 1.2 million (Rees, 1982) arising originally and continually from the domestic pet population, now estimated to be 6.8 million cats (PFMA, 1985).

Some people, often elderly or lonely, provide food for feral cats on a regular basis, so encouraging the coalescence of groups of cats; but pest problems may then be caused to other local inhabitants less enamoured of their feline neighbours (UFAW, 1981). Large numbers of cats accumulating in an area naturally cause a concentration of natural cat behaviour patterns, many of which are unsociable and unacceptable (Table 24.1). Pungent territorial spraying by toms, nocturnal noise from cats fighting or calling by oestrus females are the usual reasons behind feral cats becoming pests and, along with fears of zoonoses and fleas, are the usual reasons behind demands for control. Cat lovers and feeders invariably find themselves unable to cope with the rising numbers of charges and also seek to control the size of their colonies. Unspayed queens may produce up to three litters of kittens per year, most of which would die if left alone. However, the symptoms of feline infectious enteritis and feline influenza are most unpleasant, and the sight of many kittens suffering from nasal discharge, emaciation and mucus-filled eyes, coupled with the ever-rising costs of feeding and caring for a growing colony, force even ardent cat lovers to seek control measures.

For many urban pests attempts at population control are usually directed through various means of culling. Control of cats by shooting and poisoning is almost certainly illegal so long as cats are still afforded protection as domestic animals under the Protection of Animals Act,

Table 24.1 Reasons for complaint about feral cat populations

Smell (especially territorial spraying by toms)
Noise (queens calling, fights)
Health risks (fleas, zoonoses)
Numbers
Distress (sight of natural mortality occurring)

England (1911) and Scotland (1912), and the Cruelty to Animals Act (1986). Control by trapping and humane destruction of cats using chloroform or the administration of a lethal injection by a veterinary surgeon is legal and practised widely by pest control operators. The classification 'feral' has yet to be tested legally, but even when landowners have ordered control measures using legal methods, public sympathy is invariably in favour of the feral cat, because he appears identical to his pet counterpart. Strong protective emotions are aroused in most animal lovers, and even in less-ardent pet keepers, and culling of pest feral cats is often difficult and dangerous for pest control operators themselves (Neville, 1987). Even when cats have been trapped and destroyed, temporary abatement of numbers and problems is all that is ever achieved as long as food and shelter remain. The high population of stray, lost or abandoned cats in urban areas soon moves in to fill the vacuum left by those destroyed and recolonizes the vacant area, at the same time refilling it with the associated feline problems. Need for control is then required once again. In London this vacuum-filling process seems to take 3–6 months, depending on the time of year.

Whilst most pest controllers still try to eradicate pest cat colonies without success, alternative more biologically based methods of control are being explored. In the 1970s in Denmark, contraceptives were used in an attempt to control both cat numbers and associated behaviour problems. Research in the UK using megestrol acetate demonstrated the difficulties of this approach (Remfry, 1978). Individual dose rates have to be guessed and treatment begun when queens are in anoestrus, a rather short period in the winter for the true feral cat, the reproductive cycle of which is governed by temperature and day length change. Additionally, each cat has to be predictable in its behaviour to receive its oral dose at the right time, and feeders have to be able to distinguish each of their charges. In London, black and black and white cats predominate in colonies and even their feeders are often unable to identify individuals reliably. The nuisance behaviours of toms, while reduced by the drug, were not alleviated entirely and remain as potential causes of complaint. Trapping and surgical neutering was proposed as a permanent, more effective and humane alternative, and has subsequently proved much safer and more reliable in its control of both cat numbers and problem behaviours of both toms and queens.

A neutering programme entails trapping an entire colony (Table 24.2). Sick and aged individuals are culled and friendlier kittens or adults (perhaps stray rather than feral) are found homes to reduce the total numbers. Healthy young adults are neutered by a veterinary surgeon; toms by castration and queens by ovariohysterectromy. Both procedures alter the behaviour of cats after surgery and their return to site, in a

Table 24.2 Neutering programme

Trap	⟶	Euthanasia
↓	↘	
Neuter		Tame and home
Identify (ear tip)		
Treat (vaccinate/de-flea)		
Convalesce		
Release to site		
Aftercare/management		

manner which alleviates all the usual sources of complaint against feral cats. Queens obviously no longer come into season and so do not call to attract toms, and they no longer produce kittens, so removing much of the welfare complaint concerning the springtime boom of sick kittens. As patrolling and marking territory by spraying is usually an androgen-stimulated behaviour in toms, castration reliably reduces the size of their range and decreases the frequency of spraying. The spray also becomes far less pungent after castration. Male cats fight less, so reducing complaints about noise, and they generally seem to become more tolerant of other cats and live the bulk of their lives in a much smaller area, providing food supplies are adequate (Neville and Remfry, 1984).

Neutered groups of cats will defend resources of food and possibly shelter in their territory and deter immigrants, provided some level of the need to compete is maintained by providing only enough food for the colony and no more. Colonies remain stable (see Figure 24.1) over many years provided the occasional successful immigrant is trapped and neutered or culled before it breeds. Hence a properly supervised neutered colony can provide better long-term control of cat problems compared with repeated culling of an ever-reforming colony. A single cull can only provide a permanent cure if all sources of attraction for cats of food and shelter are removed from the territory – in practice almost impossible to achieve.

Since its original introduction in the 1970s neutering has met with widespread acceptance as a method of control for feral cat problems, and most urban veterinary practices will have been called on to neuter feral cats for local animal welfare groups. The veterinary surgical procedures have been described in UFAW, and the identification of neutered cats by the removal of the tip of the left ear of the neutered cat whilst under anaesthetic is also published (Cuffe *et al.*, 1983), (UFAW, 1982). Cats are trapped in specially designed cage traps designed for use with squeeze back cages which preclude the need for trappers or veterinary surgeons to handle conscious cats.

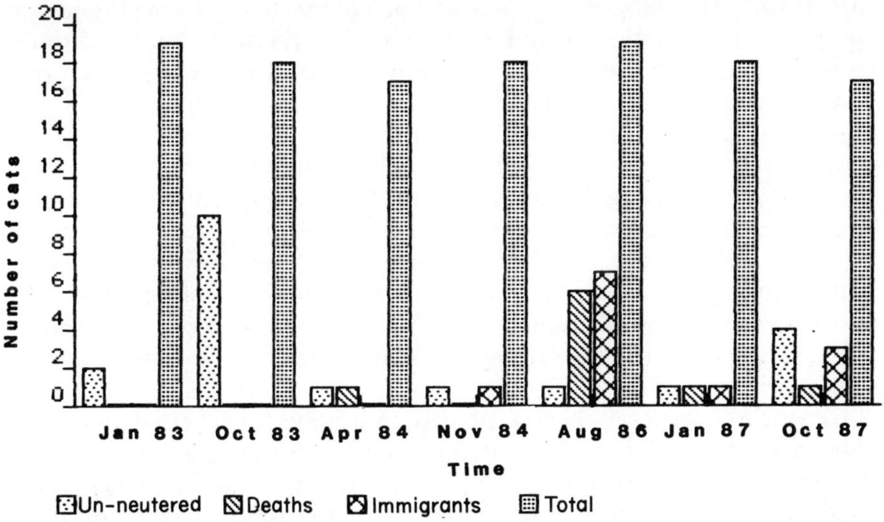

Figure 24.1 Battersea feral cat colony: population data since initial neutering programme in January 1983.

Cats are anaesthetized using intramuscular injection administered through the bars of the cage. After surgery cats are held for 24 to 48 hours for convalescence and then returned to their site (UFAW, 1982). Welfare societies and the Royal College of Veterinary Surgeons recommend the method provided the long-term welfare of cats returned is assured, minimally this means that they must continue to be fed. Neutering has been adopted as the method of choice for controlling feral cats by a number of London boroughs and is also used by many more up-to-date pest control operatives.

Research by the Universities Federation for Animal Welfare (UFAW) in the 1970s and 1980s demonstrated the cost effectiveness of such a control programme in the long term compared with repeated culling programmes (Neville, 1983). At one London colony the costs of trapping, transportation and surgery were nearly double the estimated costs of a programme of trapping and humane destruction. However, the maintenance cost over the next 4 years has been minimal and restricted to the trapping and neutering of a few successful immigrants. This contrasts with the need to repeat the culling process once or twice per year at similar costs to the original programme. The point at which neutering schemes become cost effective seems to be at about 1–2 years after the initial programme.

The effectiveness of the approach may be illustrated by one selected case study. The progress of a feral cat colony living in semi-derelict garages on a housing estate in Battersea has been closely monitored since an initial neutering programme was carried out in 1983. The population, and proportion of cats unneutered at the times of inspection is shown in Figure 24.1. The data show that the colony has remained stable throughout the five year period. The number of complaints received by the council concerning this colony has declined dramatically, and council employees too have now 'adopted' the cats as their own. The initial neutering programme has been supported by a six-monthly inspection, with trapping and neutering or destruction of any new immigrants. The maximum number of immigrants in a 6-month period was seven. However, since 1983 only one litter has been born at this colony (Neville, 1983, 1987).

Despite the general success of this technique however, it may not always be appropriate and repeated culling of reforming colonies may be the only way to deal with certain problems, for example, where cats coalesce around sensitive areas such as in the food industry or near hospitals. Equally, a neutering project to control feral cats in tourist resorts in the British Virgin Islands in 1986 and 1987 could only be described as partially successful. Although numbers and the usual cat problems were successfully controlled, the chief problem of cats defaecating on hotel roofs, which are used to direct rainfall to storage tanks to be used subsequently for drinking, was not solved. The problem was reduced by the general reduction in the number of cats, but in this instance even one cat, neutered or entire, is one too many. The hotel management rightly decided that in future all cats caught should be culled, and trapping should be carried out continuously.

However, in general and in less-specific circumstances, neutering has proved extremely successful both at managing feral cat populations and their associated behaviour problems without the distress of killing an animal which is acceptable and often encouraged by many people. Indeed the continued presence of a stable healthy colony affords great social and sociological benefit to many in our urban communities. Cat-neutering schemes have been set up and are being maintained successfully in urban areas worldwide and in tourist resorts in countries as culturally diverse as Kenya, Greece, USA, Tunisia, South Africa and Israel. Hotel managers in these countries have found the idea particularly attractive as it enables them to control cat problems effectively to satisfy those guests who dislike unwelcome aspects of cat behaviour, yet in a manner which is acceptable to the many (usually North European) cat-admiring tourists. The continued presence of a few cats as ratters is also valued in many countries.

Neutering of feral cats is therefore perhaps one of the first, now most widespread and best examples of a more enlightened and effective approach to pest control which takes account of the animals biology and behaviour.

REFERENCES

Cuffe, D.J.C., Eachus, J.E., Jackson, O.F., Neville, P.F. and Remfry, J.C. (1983) Ear tipping for the identification of neutered feral cats. *Vet. Rec.*, **112**, 129.

Neville, P.F. (1983) Humane control of an urban cat colony. *Int. Pest Control.*, **25**, 144–5, 152.

Neville, P.F. (1987) Walk on the wild side. *Bull. Feline Advisory Bureau*, **24**, 12–16.

Neville, P.F. and Remfry, J.C. (1984) The effect of neutering on two groups of feral cats. *Vet. Rec.*, **114**, 447–50.

PFMA (1985) *Profile 2*. Pet Food Manufacturers Association, 6 Catherine Street, London WC2B 5JJ.

Rees, P. (1982) The ecological distribution of feral cats and the effects of neutering a hospital colony. PhD Thesis. University of Bradford, Bradford, Yorkshire.

Remfry, J.C. (1978) Control of feral cat populations by long term administration of megestrol acetate. *Vet. Rec.*, **103**, 403–4.

UFAW (1981) The ecology and control of feral cats. Universities Federation for Animal Welfare, 8 Hamilton Close, South Mimms, Potters Bar, Hertfordshire EN6 8QD, UK.

UFAW (1982) *Feral Cats: Suggestions for Control*. Universities Federation for Animal Welfare, 8 Hamilton Close, South Mimms, Potters Bar, Hertfordshire EN6 8QD, UK.

Index